AF704492

When the People Bubble POPs
Man's Evil and Deadly
OVERPOPULATION

Published by World Audience, Inc.
(www.worldaudience.org)
303 Park Avenue South, Suite 1440
New York, NY 10010-3657
Phone (646) 620-7406; Fax (646) 620-7406
info@worldaudience.org

ISBN 978-1-935444-91-6

Edited by Kyle David Torke

Cover design and layout by Christopher Taylor

World Audience (www.worldaudience.org) is a global consortium of artists and writers, producing quality books and the literary journal *audience* and *The audience Review*. Our periodicals and books are edited by M. Stefan Strozier and assistant editors and designed by Christopher Taylor. Please submit your stories, poems, paintings, photography, or artwork: submissions@worldaudience.org. Inquire about being a theater reviewer: theatre@worldaudience.org. Thank you.

TABLE OF CONTENTS

Toynbee said that civilization's "Time of Trouble"
Means it'll soon perish as a mere pile of rubble—
A most fitting reward
If it also ignored
The impending loud **POP** of its huge "People-Bubble."

Introduction

This is a very unusual book, which it is meant to be. Parts of it, like fiction, would seem to overstep one's conventional idea of ordinary propriety as well as limits of decently fanciful human imagination. But some readers will persevere out of piqued curiosity underscored by an uneasy feeling that the book's message, no matter how outlandish it may at first strike one to be, somehow seems relevant and important to one's own existence. Others may choose to read on merely because they find the book's pretentiously farcical seriousness somewhat amusing.

Nevertheless, I trust that every reader will in the end conclude that such judgments should be of little concern and that what really matters is reflection on its dour message, which must be pondered by every sentient person of every color and race everywhere on this ailing earth. To many, the message may be unconvincing, ludicrous, and totally incapable of provoking action on any of the extraordinary and weighty, even revolutionary, corrective or preventive measures called for. Therefore, it is imperative that this book's message be deeply and permanently ingrained in human consciousness—that this book *at least be read*. And, for that alone, I thank you.

The fundamental problem is, and always was, *global*. One may well ask why this issue isn't discussed and appreciated. Perhaps the universality of the topic and its portentous consequences have become the ultimate taboo in all segments of every society. So let's begin with the basics and take a brief, clear, and honest look at the gradual and at times agonizing evolution of human terrestrial life as most of us comprehend it from prehistory to the present—embellished with knowledge and opinions of respected scholars who obviously should know, or perhaps deduce and guess, more from scientific evidence than laypersons can; and then let's take a similar look at our present and the likely fate in store for it in the near, if not immediate, future.

Eons ago, this was a still smoldering hot, plastic, primeval world of copiously flowing lava; of recently filled, probably less salty, oceans; of lush verdant landscapes covered with a thick green blanket of tropical forests, bushes, and giant fern-like plants; of all kinds of weird, heretofore unseen and unimaginable giant and tiny creatures, now long extinct. Then gradually there appeared in this splendid, gigantic diorama of Nature an extraordinary creature whose behavior compels me and many others to conclude that humans are Nature's sole mistake. Perhaps that same charge should be put to a supposedly omnipotent God endowed with obviously flawed omniscience which gave rise to this far from perfect image of Himself.

However, the latter is well-exemplified by our own flawed image of a cave man depicted as a comic book character named Alley Oop who lives with his mate, Oona. One must wonder whether as many of their scanty cohorts then would have honestly and enthusiastically agreed with the almost unanimous, often specious, proclamation of the vast majority of *our* "civilized" human ingenuity in development of science and technology that has greatly expanded our life span to thin out and mollify the episodes of tribulation, allowing much more of the "precious" time to be taken up with the joys (often more debasing than edifying) that collectively sustain the "gift" myth. Such is the power of human delusion that has been rationalized with the concocted concept of an ethereal and supposedly indomitable "will to live" that in present circumstances will almost certainly and soon be lethal on a globally grand scale.

Early human's condensed existence meant constant struggle merely to survive in the natural, brutal, but exquisitely balanced panorama that was unending, ugly, usually bloody, and short—an experience that undoubtedly inured the cave man to death, chillingly omnipresent and always imminent. Also probably doubtless was the only fear he had of death as concisely defined by the 16th century philosopher, Francis Bacon: "I do not believe that anyone fears to be dead, but only the stroke of death."[1]

Thus early humans probably feared only *how* they would die, how violent and bloody it might be. When scrounging for food, they had to risk dangerous competition with giant herbivores and carnivores, relying on a pathetic armamentarium of bare hands, clubs, crude spears, and bows and arrows in conjunction with their great and unique natural advantages of superior manual and digital dexterity coupled with a relatively superior intellect to complement their animal instincts. This enabled them initially to dominate other mammals and eventually all biotic life and led humans ultimately and foolishly to consider the possibility of challenging the hegemony of Nature itself. This has subtly and probably involuntarily enmeshed us in a fateful, titanic "war" with Nature, now evolved into a super contest not for mere dominance or control, but because of humans' arrogance, obstinacy, and perpetual folly, "for all the chips," for survival—a hegemonic battle *to the death*!

Well acquainted with basic human nature, one can easily guess who the victor will be. For even a tattered and decimated Nature *will* survive; but human survival without Nature is unthinkable. Nevertheless, our colossal hubris

---

[1] Bacon, Francis: www.dictionaryquotes.com

and the apparently predominant and often labile evil streak in our nature compels us to overlook the restraint that was plain common sense to Francis Bacon: "Nature to be commanded must be obeyed."[2] And that serious, if not fatal, character flaw prevents humans from desisting this reckless and futile "progress" toward *certain species suicide,* taking with them much of our severely wounded adversary's coveted but now inutile and to us meaningless wealth.

There you have it: the theme of this book, the "Ueber-War" in which all of humanity is now foolishly engaged with an outcome all can envision but which only the very young now alive may actually experience a bit of before their own ultimate ignominious demise. The full brunt of that monumental tragedy will be reserved for your descendents if their lineage is unlucky enough to survive a few centuries hence; and if they do, they may rightly curse your memory for your appalling lack of insight or for your indifference: how little it will seem to them whether or not we cared.

---

[2] Peter, Laurence J.: *Peter's Quotations.* William Morrow & Sons, NY, 1977, 348.

Chapter 1

Constituency of the Mega-Problem

Almost all parturition on earth
Bodes a life that's a joke without mirth;
So, fowl yolk, or placenta,
Each afterbirth meant a
Foul yoke to be borne after birth.

Human's nearest ancestor, Homo erectus, is said to have first appeared on earth about two million years ago in Africa and later migrated to Europe. With development of his brain during the following eon, he evolved into *Homo sapiens*; and after more thousands of centuries, he finally became the progenitor of *Homo sapiens sapiens*, the species of which all modern human beings are members. As populations increased slowly, and more infants survived, the first groups formed were individual families that eventually coalesced into scattered small communities. These soon learned that mutual cooperation provided more security; and thus communities naturally developed into ever larger groups called villages, towns, and, of course, eventually with the birth of civilization, to what are called cities.

Evidence from archeological digs in the Pleistocene era suggests that about 15-20% of all live births at that time had to be eliminated, often by infanticide (i.e., purposely killing the infants up to one year of age) or neonaticide (killing newborn infants immediately or very soon after birth)—apparently an instinctive necessity to guarantee enough sustenance for the protection and survival of the adults and the whole group. On the other hand, children were considered to be valuable if pestilence and war with neighboring groups did not suffice to limit population growth.[3] The same was true for nomadic groups where having many children was not considered a burden, but rather a blessing.

In our time, the noted ethicist Joseph Fletcher considered infanticide to be a justified and honorable practice, performed out of "loving concern" without denying an infant's personhood.[4] Infanticide was common in ancient Greece, Rome, Carthage, and China. It was also practiced in Europe until the 19th century where it included the practice of "overlaying" (smothering) an infant sharing a bed with a mother or father, and the abandonment of unwanted infants by depositing them into sewers in larger cities or in foundling hospitals established to handle the increasing use of the practice—and most of the confined infants usually perished in those foundling hospitals anyway.[5]

As birth control spread among upper class Romans, Caesar in 59 BC rewarded women who had many children; and forty years later, Augustus did the same (a tactic sometimes used in a few of our modern societies with equal futility). At that time, Romans practiced some form of limiting population, such as contraception, abortion, or infanticide. Girls or illegitimate or deformed children were

---

[3] Sumner, Wm. Graham: *Folkways.* Dover Publications, Inc., NY, 1906, 308.

[4] Fletcher, Joseph: *Playing God in the Nursery.* W. W. Norton Co., NY, 1985, 229.

[5] *Encyclopedia Britannica, 15th ed., vol. 25.* 2005, 1036.

killed by exposure to the elements.[6] Pagan society was being decimated by abortion and infanticide. Christians considered infanticide the equivalent to murder and opposed the practice. In Greece, Plato advocated the killing of deformed or "inferior" infants and children to improve society; and almost all the coeval philosophers approved of family limitation. A father in Sparta had the right to kill any apparently defective child by throwing it off a cliff onto jagged rocks below.[7]

In Europe, Christianity stopped infanticide and lessened the incidence of abortion. Mohammed banned the practice for Muslims, and the Koran prohibited it. But it was still relatively prevalent in 12th century France where Charlemagne decreed that exposed children should be the slaves of those who rescued and reared them. Around 1190, a French monk founded an institution to protect and educate orphans—probably the prototype of later foundling hospitals in several European countries.

In the 4th and early 5th centuries, economic and demographic crises were transforming Rome into an oppressive totalitarian state with a falling birth rate, forcing the state to protect continuity of human life by banning contraception, abortion, infanticide, and abandonment of infants. Concurrently repressive measures against suicide were being put into place. The punishment for suicide was made harsher than that for murder (which was only fined); voluntary suicide for martyrdom disappeared.

Among Anglo-Saxons, suicides were supposedly due to insanity or to being possessed, and it was not punished. From the 4th to the 10th centuries, suicide in Western

---

[6] Durant, Will: *Caesar and Christ*. Simon & Schuster, NY, 1944, 364.

[7] Durant, Will: *The Life of Greece*. Simon & Schuster, NY, 1939, 187.

Europe was totally banned by Christianity despite the prevailing climate of bloody violence and the scorn for human life—and mainly to alleviate the pressing need to maintain and increase the labor supply.[8]

Some 14th century sentimental jurists recognized the grandeur of suicides in classical antiquity and called for mitigation of the punishment for any suicide due to the distaste for life, such as the suicides of Cato, Diogenes, and Zeno. Ordinary suicide in the Middle Ages was considered to be of the petty, egotistical kind, carried out by laborers, serfs, peasants, and craftsmen—like cowards seeking to avoid the tribulations of life and thus adding to the low esteem in which the act was held[9] and which was later incorporated into the law. Martin Luther considered it to be murder committed by the devil. In France, after 1690, the increasing number of prisoner suicides alarmed the government, which tried to characterize them as "accidents" in its jails; and that led to demands that penalties for ordinary suicides be applied and enforced.[10]

Suicide because of distaste for life and for the absurdity and worthlessness of existence (so-called philosophical suicides) were rare and dubious in the late 1700s, but they frequently occur in more recent times. The pessimistic philosopher, Arthur Schopenhauer, expressed it thusly: "How shall a man be proud, when his conception is a crime, his birth a penalty, his life a labor, and death a necessity?[11] ...The nature of life throughout presents itself

---

[8] Ibid., 81.

[9] Minois, George: *History of Suicide*. Johns Hopkins Univ. Press, Baltimore, 1999, 29-31.

[10] Ibid., 200-202.

[11] Saunders, Bailey T.: *Complete Essays of Schopenhauer*. Wiley Books, Inc., Abridg. of vols. I-VI, NY, 1942, 3-Book VI.

to us as intended and calculated to awaken the conviction that nothing at all is worth our striving, our efforts and struggles; that all good things are vanity, the world in all its ends bankrupt, and life a business which does not cover expenses; that the loveliness of life is a lie, and that the greatest boon of all is death…To be happy, one must be as ignorant as youth."[12]

Several philosophers had a similar opinion. For Nietzsche, "The best of all is unobtainable—not to be born, to be nothing."[13] Poet Heinrich Heine summed it up briefly: "Sleep is lovely, death is better still, not to have been born is of course the miracle."[14]

According to the eminent socio-anthropologist Sumner, the humanitarianism, which arose from the 15th century mores, was due to the increased power of humans over Nature through the acquisition of new land and by advance in the arts. And it offered a different outlook on voluntary death.[14] On the other hand, I would agree more with the opinion of the noted Harvard anthropologist Earnest A. Hooton that "through stupidity and sentimentality humanitarianism utilizes knowledge to aggravate the evil evolutionary status of man, and to help foster the social cataclysm which bids fair to destroy civilization." It does this by promulgating "such ideas that the strong should protect the weak, that human life is sacred, that each individual should be allowed to pursue happiness, that those who have should share with those

---

[12] Durant, Will: *The Story of Philosophy*. Garden City Publishing Co., Inc., Garden City, NJ, 1927, 356.

[13] Ibid., quotes Silenus, 442.

[14] Sumner, Wm. Graham: *Folkways*. Dover Publications, Inc., NY, 1906, 39.

who have not, and that equality of opportunity should be accorded to all members of a society. Such an altruistic scheme...is practicable, if at all, only in a society whose human constituents are individually and collectively of relatively high intelligence. It is contrary to nature and to evolution, and though noble, virtually superhuman."[15]

Furthermore, Hooton stated that "a policy of unenlightened humanitarianism multiplies the physically unfit and establishes population ratios in which the denominator of inferiors increases much more rapidly than the small numerator of the fit. Humanitarianism...devotes its attention to the preservation and succor of the weak through the socialization of medical science, through private and public charity, and through toleration of antisocial individuals...and their maintenance at the expense of the law-abiding."[16] However, Hooton admitted that "charity is necessary, and many of its beneficiaries are deserving. Its less encouraging result is to relieve the least capable of the population of the necessity of supporting themselves, and to permit them to live and *breed* (my emphasis) copiously...Much charity fosters human stock of inferior intelligence, which is willing to take liberty but not to grant it to others."

Let's now focus on the key element in the Mega-Problem: peoples and their nature. It is estimated that in the year 1 C.E. (Common Era) and 1 A.D. (Anno Domini) the world population was about 300 million (See Figures 1 & 2). Now plotting the population figure for various dates reveals an astonishing curve that should also be alarming. It was to be expected that the curve would be relatively flat for much of the time period along the basal horizontal time

---

[15] Hooton, Ernest A.: *Twilight of Man*. G. P. Putnam & Sons, NY, 276.

[16] Hooton, Ernest A.: *Twilight of Man*. G. P. Putnam & Sons, NY, 279.

line (the abscissa) of the graph. The graph's vertical line (the ordinate) denotes estimated world population in billions. Note that it took 1,250 years, twelve and a half centuries(!), to add one million more people to the world's estimated total. And then, in only about 250 years, *five hundred* million people were added to finally reach *one billion*. The sudden drop in about 1350 was due to the calamity of what is known as the "Black Death," which in a few years quickly wiped out almost half of the population. (It will be discussed in more detail later.) Note also that in only about one hundred fifty years, the massive loss from the plague was not merely recouped, but topped by one hundred million.

This type of curve with a rapidly bending line is called exponential or logarithmic to indicate that the value on either the ordinate or abscissa constantly changes while the other's value remains constant. This graph shows clearly and convincingly that human population is now expanding at an incredible and ultimately unsustainable rate—portending an unimaginable catastrophe fairly soon (on a geologic time scale).

As mentioned earlier, both primitive and civilized societies realized that the resources of earth's natural ecosystem would at some point impose a limit on the size of their populations in any contrived political or economic system. Taken as a whole, the earth's entire ecosystem must therefore have some kind of indefinable absolute limit for all peoples; but it's highly doubtful that it will ever be reached—for a couple of reasons: First, it's no secret that human beings are extremely reluctant to attempt extremely radical changes, no matter how necessary. Second, the noted English philosopher-astronomer Sir Fred Hoyle propounded a salutary scenario replete with extremely radical changes that entail a series of massive overpopulation probably none of which will be convincing as an absolute limit (we'll get to Hoyle's suggestion later.)

Third, Western civilization seems to be breaking down,[17] which in all likelihood eliminates any possibility of humankind ever reaching that limit. Finally, the most compelling reason is human's incurably fundamental irrationality—with reference to his incurable addiction to the cozy illusion that it's always wonderful to be alive.

We turn now to Hoyle's intriguing theories. In his opinion, Malthus was wrong in assuming that starvation would hold population at subsistence level, even though the people are not starving. It is the *organization* of such a community that ultimately becomes overloaded and collapses because it would be impossible to maintain the necessary organization if there remained only one square yard of standing room per person. Hoyle stated further that if the population were to continue to rise in accordance with the exponential rule as at present, then in a mere five centuries only one square yard of standing room per person would remain on the earth's surface; and in fifty centuries, the total mass of H. sapiens would exceed the mass of all the stars and visible galaxies!

Rhetorically, Hoyle asks, "What is going to change the exponential rule?"[18] His answer, expressed precisely and cogently, is that something entirely new is needed; and he lays it all out in detail—as discussed in the next chapter and simply summarized in Figures 1 & 2.

---

[17] Durant, Will & Ariel: *The Lessons of History*. Simon & Schuster, NY, 1968, 100.

[18] Ibid.

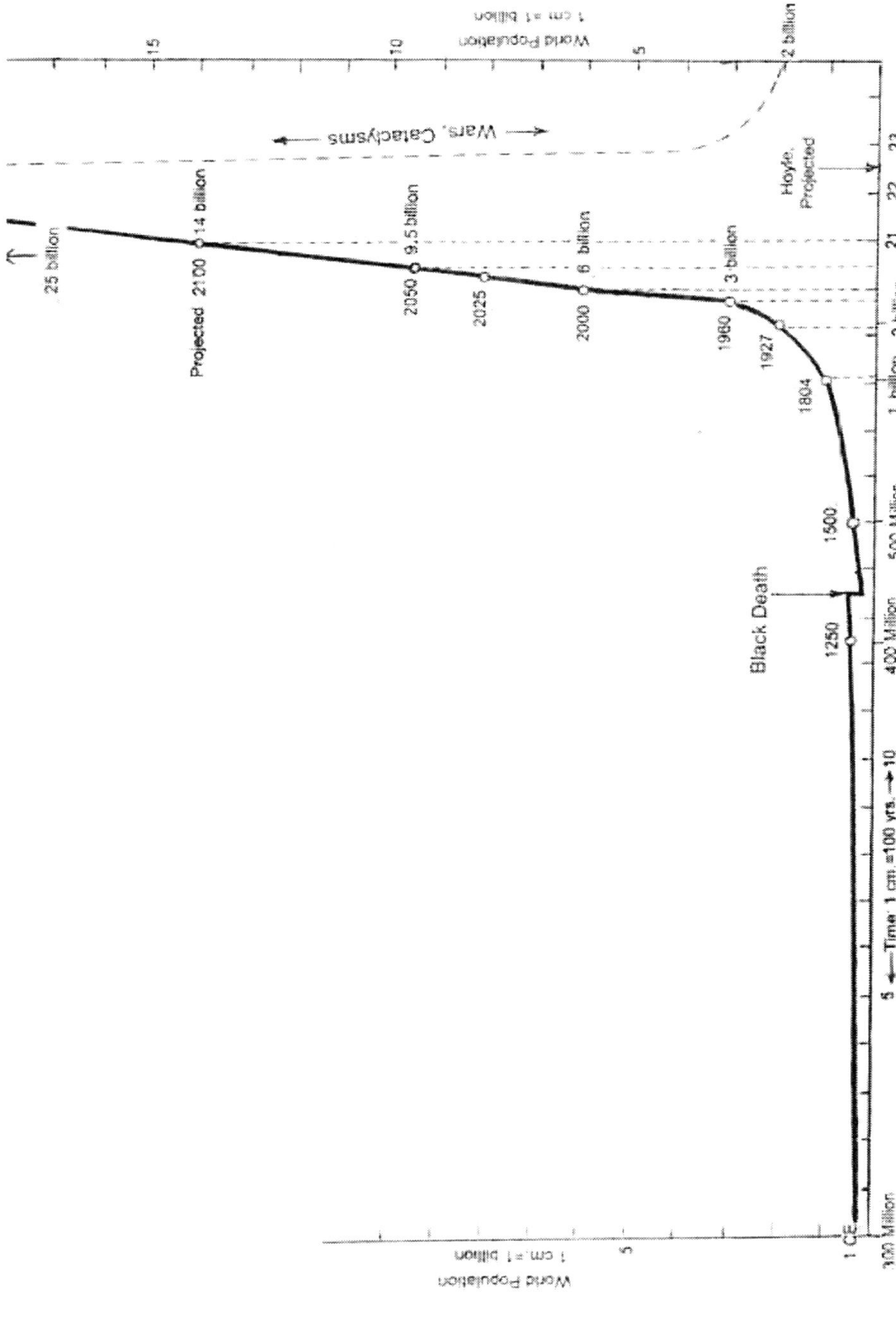

Fig. 1 - Projected cycle Graph based on Hoyle, Fred: On the Argument of Malthus. *Population & Development Rev.*, vol. 12, no. 3, Sep. 1986
Data from *The World Almanac 2009*, World Almanac Books, Pleasantville, NY, 2009, pp. 715, 716.

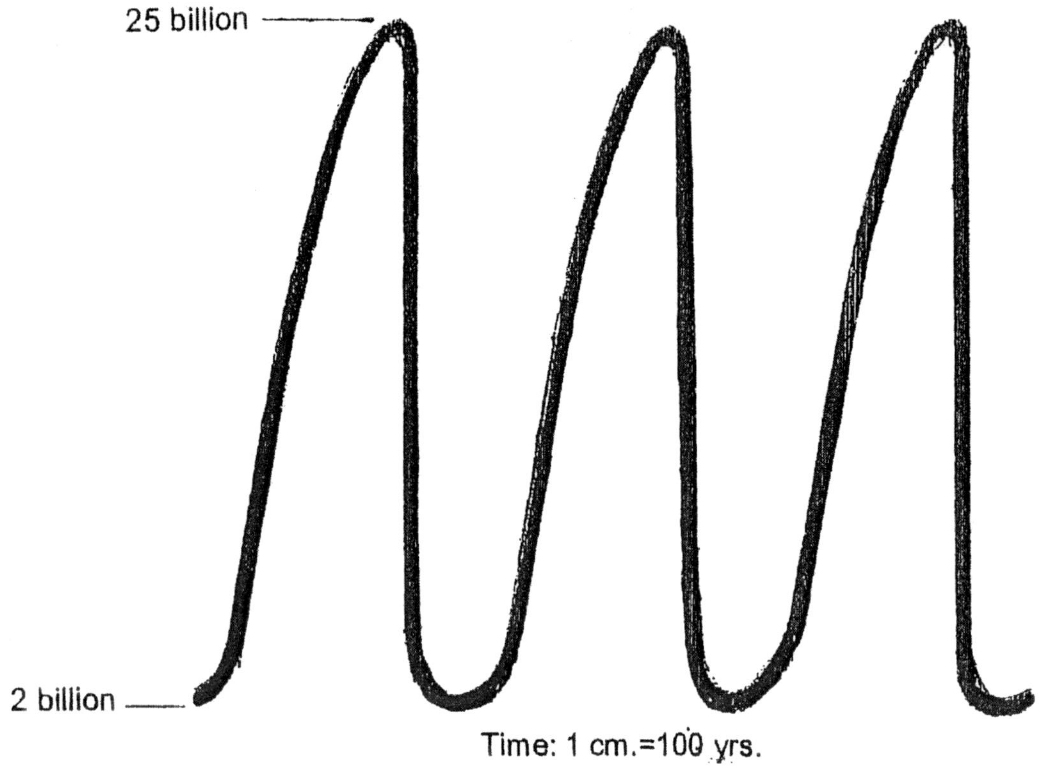

Fig. 2 — World Population

Projected cycles based on Hoyle, Fred: On the Argument of Malthus. *Population & Development Rev.*, vol. 12, no. 3, Sep. 1986

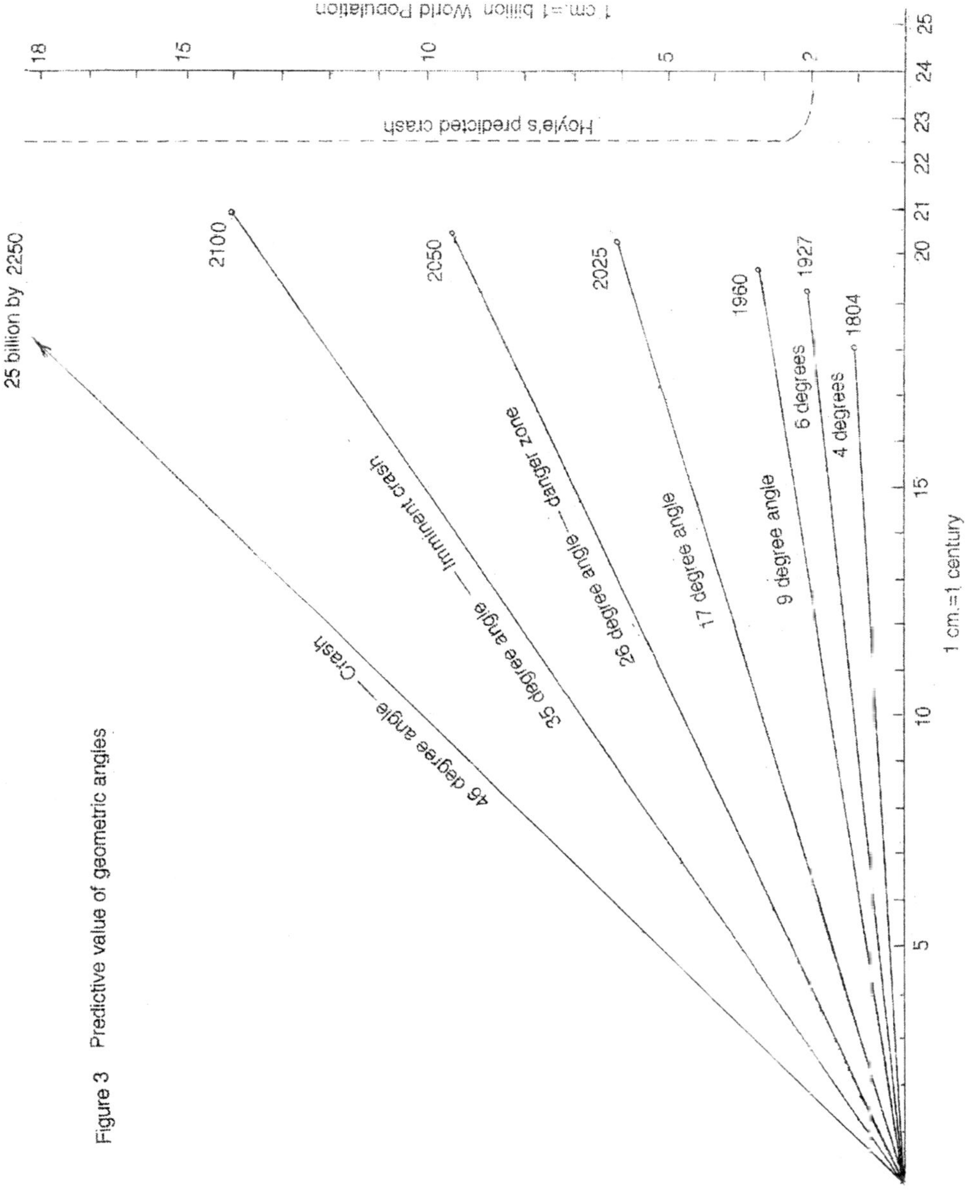

Figure 3 Predictive value of geometric angles

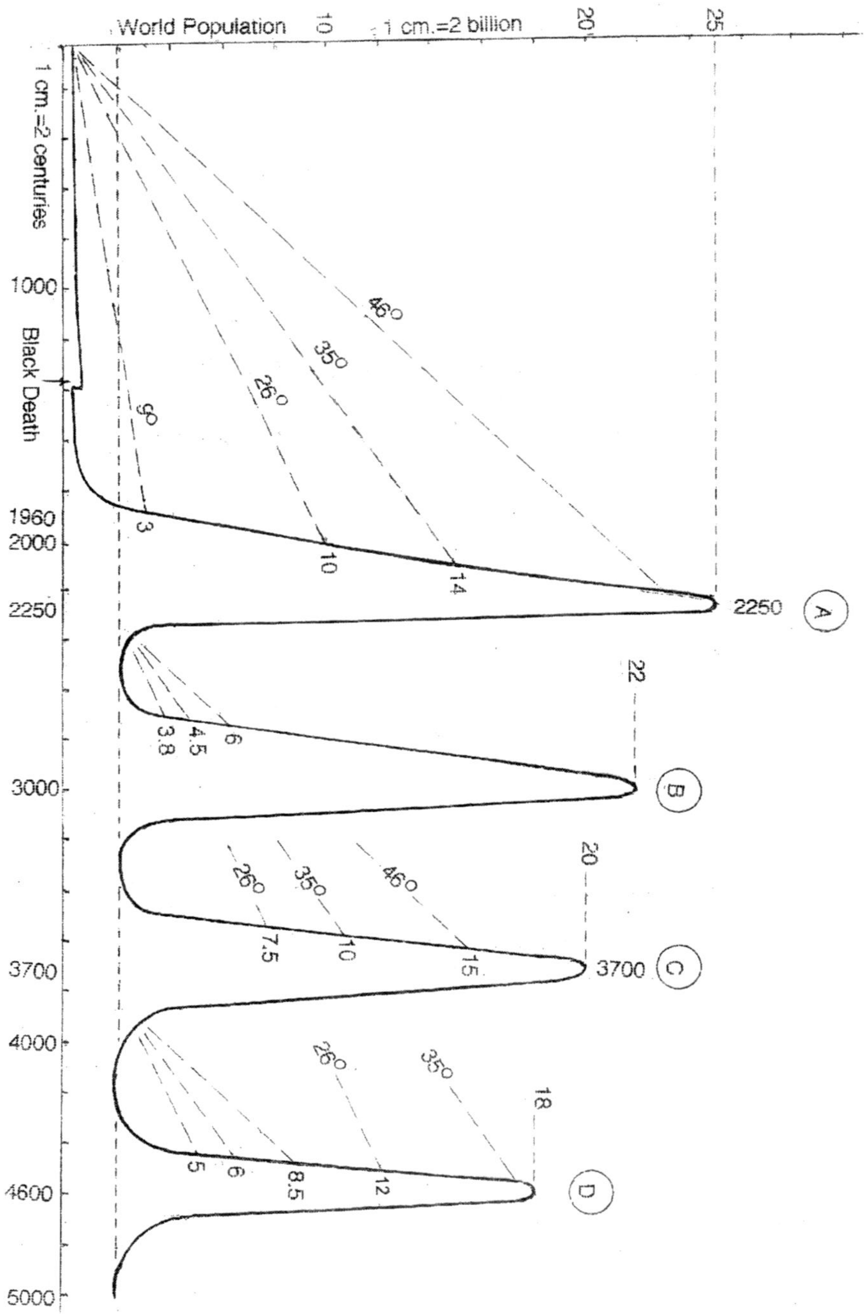

Fig.4 Projected future population

Chapter 2

Fateful Cycles of Overpopulation

With the cyclical option man cannot win,
For he'll continue committing the biological sin
Against Nature's ecology,
And his best apology
Would be his species' will to do itself in.

According to Hoyle, it is well known in the physical world that "organizational structures are subject to discontinuous change." He cited, as an example, ice melting to water and assuming a new physical organization that does not change continuously with rising temperature; there is a *single* different temperature at which the decisive change occurs. So Hoyle therefore suspected that, similarly, the rise in the population total to a certain level will disrupt human social organization.

He had no doubt that a major global war one or two centuries hence could produce the necessary discontinuous change.[19] He added, unfortunately before the possibility of a thermo-nuclear threat, that "warfare itself never significantly increases the death rate. As a cause it can't compare to the Black Death." Concomitant with the discontinuous collapse of society will be a

---

[19] Hoyle, Fred: *On the Argument of Malthus*. Population & Development Rev., 12, no. 3, Sept., 1986, 554.

discontinuous rise in the death rate because the very organization that permitted vast numbers to be fed and kept alive will have disappeared. He admitted that at this stage we are forced, to some extent, "to make an effort of imagination instead of logic…The more complex society becomes, the greater the disaster will be…similar in its effect on the population to that of the Bubonic Plague." He concluded that the drop in population and its subsequent rise…"will be like that of the plague, but the numbers will be higher." And there will be a rise in one aspect "in which the human species has shown not the slightest originality—its *excessive reproductive vigor*[20] (my emphasis). The tattered remnants will slowly reform and reestablish themselves."

He believed that the invincible "excessive vigor" will guarantee more than one repetition of the cycle (which he depicted as three cycles of equal size and shape and equidistantly separated). The most fit among the tattered remnants (already skilled in the basic arts and technological know-how, adept in means of survival, tempered by experience, and having access to the few libraries that probably will also survive the chaos) will, through a unique process of selection in what Hoyle projected may require a series of perhaps 10 or more such cycles, survive (as depicted in Figure 2). And even without mutations, that degree of selection should yield a new creature. It will still be a human being, but "perhaps of a higher, more intelligent and more sociable type"—as just another subspecies one could dub Homo sapiens sapiens sapiens.

Still, according to Hoyle, we are now in the first cycle and "have the advantage of abundance of natural, not soon replenishable, assets," such as coal, oil, metals, and

---

[20] Ibid., 551.

minerals. Their diminution with each subsequent cycle, he hoped, "will be made up by more intelligent and probably wiser beings having valuable experience and access to the information in the libraries still extant." When this happens, he assumed, that in contrast to the primitive status of life at the beginning of the first cycle depicted, there would be no likelihood of repetition of the extreme prolongation of the beginning of subsequent cycles as shown in Figure 1. He therefore theorized that the number of cycles will be determined by the amount of selection that emanates with each cycle; if their number is large, then terminal cycles should be progressively smaller and lower in amplitude until the new species succeeds in stabilizing the population level and thereby eliminating the cycles' cause. Hoyle also translated his theories into mathematical terms to explain and summarize the text using many calculus formulas in an extensive appendix.

He predicted that the duration of each cycle, especially the first one, should be about 2 to 3 centuries, and that the populations should rise to about 25 billion in the year 2250 at which level there will be a cataclysm due to war (thermonuclear?) or worldwide pestilence resulting in a precipitous drop when about 23 billion individuals will perish, leaving about two billion tattered but wiser remnants to resort to their irresistible excessive reproductive vigour (as the Englishman Hoyle spelled it) to fairly quickly start the next, hopefully abbreviated, cycle. Hoyle apparently failed to foresee, or simply discounted overpopulation per se, as a primary, direct, and sufficient cause of the precipitous drop, assigning it only an indirect role.

Most of the data used in Figure 1 were reported by experts and therefore fairly reliable. The pictorial presentation of the data was especially startling and led me to ponder the possibility of deriving from it some kind of more objective way, perhaps numerically, of enhancing its predictive potential—similar to the well known "smog index." I hit upon the idea of using triangles of "plain" geometry to obtain the useful numerical values summarized in Figure 3.

It is obvious that a straight line from the point of origin of the cycle concerned (here, of course, the first one depicted in Figure 1) to a plotted data point forms a right-angled triangle with the projected straight line as its hypotenuse. If the abscissa and ordinate that form the horizontal and vertical sides, respectively, of the triangle are equal in length, the hypotenuse will have an exactly 45-degree angular slope indicative of a too rapid rise (for that specific cycle) in population in a relatively short time—a critical danger signal. A relatively longer abscissa as the time period in Figure 1 from 1 C. E. to the year 1804 yields a very flat hypotenuse, indicating an extremely slow increase in population. A relatively longer ordinate yields a much more vertical hypotenuse, indicative of population growth at essentially an unsustainable rate and therefore a coming inescapable crash to catastrophe. These conclusions have been summarized in Figures 3 and 4, respectively, along with their appropriate predictive significances. Thus, I theorized that numbers attached to angular slopes of these virtual hypotenuses may serve as a kind of "Population Index" for a simple, easily understandable short-hand way of roughly classifying stages of population growth. Only experience will show whether or not this suggestion is worthy of further refinement or implementation.

For me, the graphs were enlightening and useful. First, an orderly concatenation of relatively closely packed cycles (A, B, C, D) probably is not tenable. A much longer period of a well controlled population level following this initial cycle (A) would be required by Nature to sufficiently replenish its exhausting resources. Based on the "Index," I would doubt the accuracy of Hoyle's guess of a restoration period lasting for three or even a few more centuries. For such a short interval, the "index" indicates that a cataclysmic world population level of only 6 billion would be reached much more rapidly in the second cycle; whereas it took 2000 years for that same level of 6 billion to be attained by an un-tattered, not unduly concerned, and only nominally threatened population basking in the apparent opulence of an intact, comparatively pristine Nature, which will *not* be the case in the succeeding second cycle that will require restoration period of many—perhaps 10 to 20 centuries of extremely *carefully* controlled population levels between 2 to 3 billion, comparable to the pattern established before the accelerated rise took hold in about 1950 (Figure 1).

It should be pointed out that between the 5th and 15th centuries of the first cycle was a period of miserable living conditions constituting what are known as "The Middle and Dark Ages." Therefore, it is quite probable that after the first cycle, a convalescent, "tattered" populace dependent on the diminished resources of a now even more weakened Nature will inevitably have to endure a protracted period that will similarly embody perhaps a much "Darker Age" with even more miserable conditions. Such a repetitive pattern of up to 10 cycles (as predicted by Hoyle) will probably require millennia of progressively lower population levels that "trigger" the "crashes" that usher in perhaps even longer post-crash periods of "convalescence" with their progressively more brutalizing Dark Ages. The final result would most likely be the completely tattered end of our species rather than its rescue and perpetuation by a culturally evolved savior in the form of Hoyle's much improved human being.

At any rate, I accept as plausible most of Hoyle's ideas regarding the population problem. His predicted year of 2250 for the maximum 25 billion mark probably is accurate. The heavy line with the strong vertical upswing in Figure 1 does indeed indicate that it should happen sometime in the 22nd century (the peak, unfortunately, off the book page), but factual population data for the 21st century will be needed to verify the line's true trajectory.

However, I do disagree on a couple of points. First, as mentioned above, an equally tattered Nature may need more than a couple of centuries to restore the resources needed by a now "improved" humankind to perpetuate a modern, more truly civilized lifestyle. Second, I am not as wont as Hoyle to discount the powerful influence of innate and often fickle human nature in trying to predict future human behavior, especially in the extraordinary and bizarre circumstances imagined nor to naively assume a sort of revolutionary, non-mutotic genetic change in so short a time span as that of a mere solitary but colossal "Black Swan"—like event that was the Black Death[21]. It would seem to make more sense simply (and much more fantastically) to assume such a likely very short-term but actually *mitotic* change as a real, probably structural effect on the specific gene(s) controlling the essence of human nature due to some kind of rare, known or undiscovered cosmic radiation. And what are the odds of *that*?

In this regard, who should say it better than the famous Harvard anthropologist Earnest Hooten: "If man wants to behave better, let him become a biologically better animal...The situation is far from helped by the fact that criminals, morons, cogenital incompetents and biological inferiors generally are not only coddled by well-intentioned busybodies, but also permitted to reproduce their kind." He said that sociological tinkering will not get anywhere until it gets better material to work on. Schopenhauer's prescription for getting it was "a true aristocracy, brought about by the...marriage of the noblest

[21] Taleb, Nasim Nicholas: *The Black Swan.* Random House, NY, 2007.

men with the cleverest and most intellectual women."[22] I believe that the "higher" type of man who might issue from Hoyle's cycles will, like most men today, fearfully shy away from highly intelligent women with such quality. Besides, most people marry blindly and ignore Nietzsche's demand that "thou shalt not only propagate thyself, but propagate thyself upward!"[23]

Hoyle seemed to have overlooked the fact that human nature fundamentally is an abstraction that most likely is totally refractory to human manipulations. Even surgery, or grotesque lobotomies of brains, didn't help. Nietzsche explained it briefly and pointedly: "The earth has a skin and that skin has diseases; one of its diseases is called man."[24] He neglected to add that it is probably incurable. Many famous, intelligent, and highly respected individuals would have agreed with him and with Einstein, who said: "It is easier to denature plutonium than to denature the evil spirit of man."[25] The evil always seems to predominate in the end, and this will be the topic for the next chapter.

---

[22] Saunders, T. Bailey: *Complete Essays of Schopenhauer*. Wiley Book Co, NY, 1942, Book VI, 41.

[23] Durant, Will: *The Story of Philosophy*. Garden City Publishing Co., Inc., NY, 1927, 462.

[24] Brussell, Eugene E.: *Webster's New Dictionary of Quotable Definitions. 2nd ed.* Prentice Hall, Englewood Cliffs, NJ, 1988, 348.

[25] Peter, Laurence J.: *Peter's Quotations*. William Morrow & Co., NY, 1977, 253.

Chapter 3

Human Nature

Perhaps hidden away in some gene
Or the brain or some gland endocrine
Is man's nature obscure
With a trait that for sure
Proves he's evil—and often quite mean.

"Human nature is a single, fixed, inherited attribute—a common property of all members of our species...Human nature is not the same from society to society or from individual to individual, nor," according to Ehrlich, "is it a permanent attribute of Homo sapiens...Genetic instructions are of great importance to our natures, but they are not destiny. Human cultural evolution can overwhelm genetic evolution." Today's human natures everywhere are diverse products of cultural evolutionary processes.[26] The famous Greek "Father of History," Herodotus, reported an unequivocal example of the power of cultural evolution, the exact origin of which I failed to find after an exhaustive library and Internet search. Here it is verbatim:

---

[26] Ehrlich, Paul R.: *Human Natures.* Shearwater Books, Washington, D. C., 2000, 7, 9, 13.

> When a child is born (to the Tausi), all its kindred sit around it in a circle and weep for the woes it will have to undergo now that it has come into the world, making mention of every ill that falls to the lot of mankind; when, on the other hand, a man has died, they bury him with laughter or rejoicing, and say that now he is free from a host of sufferings, and enjoys the completest happiness.

Here is a totally realistic, natural viewpoint on the two most fundamental concepts of existence, birth and death, that is absolutely contrary to ours based on the totally artificial viewpoint elaborated by the mythology of religion. No wonder our civilization is inundated with depression, wars, genocides, crime, lying, and teenage suicides among the untold immorality we are compelled to tolerate. And those who cannot must often find refuge in insanity or in the teenagers' way of suicide.

Those same packages—human behavior patterns and their physical foundations—are also the source of our most serious current problems. Philosopher Will Durant claims that human beings are normally equipped by Nature (i.e. heredity) with six positive and six negative instincts to meet or avoid the basic challenges of life. Each instinct generates habits with their feelings, and their totality is any human's nature. "Human evolution during recorded time," continued Durant, "has been more social rather than biological."[27] Schopenhauer emphasized that "man is the only animal which causes pain to others without any further purpose than just to cause it." This is what constitutes "the diabolical feature in his character which is

---

[27] Durant, Will & Ariel: *The Lessons of History*. Simon & Schuster, NY, 1968, 34.

so much worse than the merely animal...But it is *Schadenfreude*, a mischievous delight in the misfortune of others, which," he says, "remains the worse trait in human nature, as a feeling akin to cruelty."[28]

Many renowned individuals had the same opinion about human nature. Freud's was decidedly negative: "I have found little that is good about human beings; in my experience most of them are trash."[29] Many others agreed, such as the existentialist Jean-Paul Sartre: "Hell is other people,"[30] and the pioneer psychiatrist Karl Jung: "The only thing we have to fear on this planet is man."[31] Even the venerable daVinci: "Nothing will remain on earth, or under the earth, or in the waters which will not be persecuted, disturbed, and spoiled, and those of one country removed into another. Shouldn't the earth open and engulf them in the fissures of its vast abyss and caverns, and no longer display in the light of heaven such a cruel and horrible monster as man?"[32] But the harshest denunciation was by Schopenhauer: "Man is at bottom a savage, horrible beast. We know it if only in the business of taming and restraining him which we call civilization. Hence we are terrified if now and then his nature breaks out. Wherever

---

[28] Saunders, T. Bailey: *Complete Essays of Schopenhauer, Abridg. of vols I-VI*. Wiley Book Co., NY, 1942, Book VI 19.

[29] Berman, Louis A.: *Proverb, Wit & Widsom*. The Berkley Publishing Group, NY, 1997, 322.

[30] Ibid.

[31] Peter, Laurence J.: *Peter's Quotations*. William Morrow & Sons, NY, 1977, 199.

[32] Allen, Steve: *Meeting of Minds, fourth series*. Prometheus Books, NY, 1989, 104.

and whenever the locks and chains of law and order fall off and give place to anarchy, he shows himself for what he is."[33] And that "show" will open soon in your own "theater."

The real essence of human nature is no secret to anyone who heeds with a clear mind the lessons gained from a lifetime of merely having experienced the gauntlet of existence. Ask any Afgan, Iraqi, Jew, Armenian, Darfurian, Indian "untouchable," or veteran of militant combat (especially trench warfare or incessant bombardment) what human nature is, and odds are that you, the reader and average comfortable citizen anywhere in most of the world, wouldn't agree with their answers or with Schopenhauer's opinion that "the satisfaction of the reproductive impulse is utterly and intrinsically reprehensible because it is the strongest affirmation of the lust for life. What crime have these children committed that they should be born? If, now, we contemplate the turmoil of life, we behold all occupied with its want and misery, straining all their powers to satisfy its infinite needs and to ward off its multifarious sorrows, yet without daring to hope for anything else than simply the preservation of this tormented existence for a short span of time."[34] Isn't that how we "civilized" human beings live—and extol it as a wonderful "gift"?

A few lucky individuals have the advantage of a mind with keen insight kept clear and honed by honest education (mostly by self) and reinforced with the wisdom of aging. Karl Jaspers was one of them: "Everyone thinks he is free. Yet it seems as though freedom has become empty.

---

[33] Saunders, T. Bailey: *Complete Essays of Schopenhauer*. Wiley Book Co., NY, 1942, Book VI, 14.

[34] Durant, Will: *The Story of Philosophy*. Garden City Publishing Co., Inc., Garden City, NY, 1927, 370.

This emptiness evokes a yearning for dependence, for being led. They flock to those who in the name of liberty will rid them of liberty." Jaspers then said what everyone in every so-called democracy should never forget: "Totalitarianism is the universal, terrible threat of the future of mankind in a mass order. It is not easy to see through totalitarianism, and so (it) achieves reality, while the victims go on existing as a mass of living corpses. An absolute dictatorship, once in the saddle, *can no longer be unseated from within*!"[35] (emphasis mine). We are obligated to pay attention to his words, not only because of his impeccable academic and professional credentials as psychiatrist and philosopher, but also because he lived in Germany through the Nazi terror to the end of World War Two. And who else should know how better to advise us?

Yet, America is rapidly sliding in that direction, and Ayn Rand called it to our attention: "There is, however, one difference between the type of fascism toward which we are drifting and the type which ravaged European countries. Ours is not a militant kind of fascism, not an organized movement of shrill demagogues, bloody thugs, and juvenile delinquents. Ours is a tired, worn, cynical fascism by default, more like the collapse of a lethargic body slowly eaten by internal corruption."[36]

Jaspers and Rand are not alone. Here, still alive in America, is the eminent lawyer Gerry Spence who, with a clear mind and insight, tells us: "Always we are told the master's interests are ours. As a consequence the master identifies our enemies for us—whom we should hate and

---

[35] Jaspers, Karl: *Philosophy and the World*. Regnery Gateway, Washington, D. C., 1989, 71.

[36] Rand, Ayn: *Capitalism: the Unknown Ideal*. New American Library, NY, 1967, 220.

against whom we should send our children to fight and die." And he concludes that "profit, not enlightenment; war, not peace; and violence, not beauty are the nourishment most often fed to us by the media...In the end, network producers answer not to the people, not to Congress, not to a democratic system, not to truth or justice, but to the demands of the corporate owners and their corporate advertisers."[37]

Who are these cryptic, behind the scene, devilish string-pullers whose psychic tyranny depends on "dumbing-down" our constantly benumbed minds—most tragically of the youth? A few notable individuals knew of their existence, but, like everyone else, could not identify them. And the hidden tyrants will make sure that we never will.

Now, back to the Mega-Problem. Any realistic gambler would place his bet on the probability that the "Homo sapiens sapiens sapiens" who is expected to germinate from Hoyle's theorized population cycles will have the same nature as his progenitor—or the same proportion of evil in *its* nature. Then why bother with such radical and ultimately futile efforts to perpetuate the existence of an agent, a human, who will almost certainly foment unnecessary and unwanted terrestrial misery?

This same question should be put to misguided medical researchers today who rather foolishly are attempting to make real the vain, childish, and masochistic hope of physical human immortality. Perhaps these researchers would not have embarked on such folly had they been aware of results of a poll taken a couple of decades ago somewhere in Europe, I believe, that revealed that the vast majority of the public, young and old alike, would not choose to live forever. Yet a few researchers are

---

[37] Spence, Gerry: *Give Me Liberty*. St. Martin's Press, NY, 1998, 234,235.

firmly committed to pursuit of that goal as the so-called cutting edge of medicine (and we all know that cuts can be lethal).

For example, Dr. Michael Fossel at Michigan State University feels confident that "we can probably extend the maximum life span of humans well into the several centuries range, and perhaps beyond that."[38] Did he ever seriously consider the ultimate purpose and probable consequences of such an endeavor: that if entirely successful, it would nullify any really serious purpose for human reproduction, or, as one critic put it, would make children obsolete?

A microbiologist at the University of California in San Francisco disagrees with Fossel: "I don't want to see any anti-aging research being done because I think that anti-aging research is like doing research on anti-gravity"[39] And, a biogerontologist at Cambridge, England, concedes: "We certainly don't have anything to reverse the aging process, and I'm not very optimistic about anything to actually even slow it down very much…The rate at which you are killed for reasons other than aging determines how much effort it is worth putting into not dying of aging."[40] For me, emphatically, NONE! It calls to mind the recent foolhardy, very unpopular and wasteful cryogenic aim to freeze and preserve one's fresh corpse in liquid nitrogen for possible reanimation when some kind of "cure" is discovered in the distant future. What a colossal fear of extinction!

---

[38] Brown, David Jay: *Mavericks of Medicine*. Smart Publications, Petaluma, CA, 191.

[39] Ibid., 167.

[40] Ibid., 213, 227.

In this rather frivolous controversy, I obviously side with the dissenters and would iterate the opinions of a couple of deep thinkers in the past—such as the famous lawyer, Clarence Darrow: "The origin of the absurd idea of immortal life is easy to discover; it is kept alive by hope and fear, by childish faith, and by cowardice." Anatole France summed it up whimsically and simply: "The average man who doesn't know what to do with his life wants another one lasting forever." And even simpler and more cogently by Susan Ertz, "Millions long for immortality who do not know what to do with themselves on a rainy Sunday afternoon."

Even if such research should by extremely rare chance prove to be successful, it will lead to nothing but massively increased human misery and to the unstoppable degradation of an already ailing Nature. It is a terribly misguided effort that highlights the researchers' egregious lack of profound philosophical reflection on the myriad, potentially deleterious human and geological consequences of their appallingly quixotic aim.

Chapter 4

Civilization

From its start, the name "civilization"
Tells a lie through its own appellation
Because "civil" it's not,
Since it fostered the rot
Of man's fatally huge population.

"Civilizations have been built up principally because the members of the societies which have produced them have been made to work—to go on performing manual labor beyond the point where they have enough food to go around," said E. A. Hooton in his *Twilight of Man*. "And," he continued, "on the other hand it is perfectly natural for man to try to get somebody or something to do work for him...So our ancestors succeeded in inventing tools and...in domesticating animals and putting them to work. But civilization really began to accelerate only when the more clever and more powerful individuals contrived to hold the noses of their fellows to the grindstone...The most significant difference between the savage and the civilized man," Hooton said, "is that the former is a gentleman of leisure, and the latter is a manual laborer," who, I might add, relies on promoting civilization's acceleration with ever more dazzling hi-tech marvels to keep paving its path as, the author Elbert Hubbard had the courage to say, "with tin cans."

Hooton emphasized that in the first place there must be a favorable balance between food supply and the size of the human population. If there is a negative balance, people begin to revert to instinctive behavior; the strong seize what they need, and the weak go without. Under such conditions, liberty would be almost impossible to maintain in the consequent rioting and anarchy. In addition to the lack of food, maintenance of liberty depends upon the anthropological character of the people, including race, physical features and state of physical and mental health and disease, and upon the individual and collective level of intelligence—whether fit or unfit. In our society, the puny individual suffers less physical inferiority, partly because, as mentioned earlier, humanitarianism has fostered sentiments in favor of protecting the weak and partly because of the opportunity for self-support through occupations requiring little physical exertion. Hooton admitted that it is a somewhat noble concept, but is contrary to Nature and to evolution. Furthermore, he contended that mixture of markedly different races may improve vitality. But that consequence may be an obstacle to population control because it not only improves longevity but also increases *fertility* (emphasis mine).

Historian Arnold Toynbee identified a combination of social evils, clearly indicating our decadent slide through failure to deal with specific challenges, some subtle and others not so subtle, but all thoroughly pervasive. Among them are changes in, or abandonment (and I would also add vulgarization) of, styles of language, dress, music, dancing, painting, and sculpture—"not a technical but a spiritual decline." Among these evils are the breakdown of common politeness, decency, and especially common sense at all levels—intensified by breakdown of the family structure and of secular morality with consequent rampant marital infidelity; adult addiction to the childish delights of sports spectaculars; widespread pornography in all media;

the most crass language openly uttered by both sexes; and a crippled and obsolete educational system capable only of turning out disinterested, uneducated, usually moronic youths who become ignorant components of a "brainless mass" (herd would be more descriptive) of ignorant ovine adults.[41]

Essayist Curtis White's trenchant question hits the bulls-eye when he asks, "How can intelligence function in a period dominated by the cynical and profitable will-to-stupidity?"[42] And, I might add, a period dominated by the rather artificially vague legal concept of "due process" to the almost total exclusion of the far more important and essential virtues of honesty, self-discipline, and especially *common sense.* This is amply exhibited by the several somewhat amusing small-claims court sessions on daytime television, supplemented by popular and vacuous TV "soap operas" and inane "sitcoms."

Toynbee cited and explained plausible reasons for our civilization's decline: "We have forfeited our spiritual birthright; our abandonment of our traditional artistic technique which is manifestly the consequence of some kind of spiritual breakdown in our Western civilization."[43] That forfeiture virus has, I might add, now unfortunately infected and begun to corrupt societies in every continent, even those which formerly were contemptuous of, and absolutely hostile to, the influence of much of Western culture. And the potent virulence of that incurable virus

---

[41] Hooton, Ernest A.: *Twilight of Man*. G. P. Putnam & Sons, NY, 1939, 266, 267.

[42] White, Curtis: *The Middle Mind*. Harper, San Francisco, 2004, 169.

[43] Somervell, C.: *A Study of History*. Abridgement of vols. I-VI. Oxford Univ. Press, NY, 1947.

will putrefy them all. As an example of this destructive devolution, Toynbee pointed out substitution of the Latin alphabet for the abandoned Arabic, which took place in Turkey immediately after the end of World War One. "Such abandonment of a traditional artistic style is an indication that the civilization associated with that style has long since broken down and is disintegrating—as a consequence, not a cause."

A direct and striking example of its ugly result happened in our time and continues to happen with increasing frequency and millions of innocent victims. The triumph of incredibly brutal barbarism coupled with the Muslim religion in the prior Ottoman Empire during World War One created the unspeakable horror of which the genocidally brutalized and absolutely defenseless Armenian Christian minority can attest. That first massive governmentally planned and promulgated genocide of the twentieth century can serve as a perfect example of Toynbee's conclusion that "such a society has been thrown onto the scrap heap by an iconoclastic revolt on the part of its own external and internal proletariats, in order that one or other of these insurgent forces may win a free field for bringing a new society to birth." And that was accomplished by the Albanian Ataturk's post-war radical Westernization of Turkey, including adoption of the Latin alphabet.

That diabolically planned and "scientifically" executed super-crime of murder and theft was copied almost exactly as the Jewish "Holocaust" before and during World War Two, a futile by-product of the "efficient" Nazis' brutal, genocidal, and fortunately failed attempt to foster a race of blond, blue-eyed "supermen." Thus, based on Toynbee's ratiocination, one is justified in concluding that, despite external appearances, the "civilization" encompassing today's Turkish and German societies has broken down and is disintegrating.

The eminent historian's coherent elucidation has simply and clearly defined the complex stages comprising the birth, life, and death of a civilization.[44] Of 26 known civilizations, 16 are now dead. Ours is among the 10 remaining, and all the others are threatened with annihilation or with Western assimilation. The first stage in the devolution of a civilization is called a "*time of troubles,*" followed by the second stage of a "*universal state,*" then the third stage called an "*interregnum*" (a pause or break in a series), and finally the end stage of "*breakdown*."

Toynbee and other experts in the field suggest our "time of troubles" had already arrived; and he estimates that it could cover a span of one to several centuries. Some may claim that America now is beginning to develop into a stage-two "universal state" capable of dominating the world. That is not an unreasonable claim because over half a century has elapsed since Toynbee published his theory in 1947, enough time at least to detect the beginning of such a change. If the claim proves to be true, it would further validate my contention that the "breakdown" stage posited by Toynbee would nullify as irrelevant Hoyle's posited view of a sort of "evolution" of a better human being.

By sheer coincidence, Hoyle's envisioned disastrous and precipitous drop in population, which he foresaw as probable in the year 2250, would be followed by an immediate surge of population growth into the succeeding cycle, all of which would probably coincide fairly closely with the beginning of the equally disastrous but more realistically credible "breakdown" stage of our civilization as envisioned by Toynbee. Indeed, that would be an extraordinarily rare epochal coincidence—perhaps underway in even less than a century hence!

---

[44] Ibid., 246.

Nevertheless, I sense a disquieting and unacceptable flaw in both scenarios. Are human beings condemned to experience only an agonizing and basically pointless "evolutionary" course through eons of time toward some sort of putative, unimaginable, but "desirable" goal common to either of two very different, highly abstract, cyclical mechanisms (those concerning people as posited by Hoyle and those concerning civilizations as posited by Toynbee)? When achieved, that goal in all likelihood will, in the near term, not imperil survival of the nonetheless ultimately doomed human species. Both mechanisms take cyclical evolutionary courses that can be likened to temporal sine waves of differing amplitudes and frequencies and perhaps with an undulating vector. But both mechanisms will inevitably be accompanied by a fundamentally immortal entity—Nature—that will always survive, but will suffer progressive harm while on an absolutely straight line vector edging downward. Each cycle of either mechanism will add to the progressive impoverishment of that immortal entity through stepwise diminution of its physical, geological, and biological wealth.

Now back to the matter that matters—consideration in detail of what would seem to be the sole, definitely conclusive solution to the Mega-Problem as it is distilled from even our first cycle. It may prove to be too distasteful for implementation, even repugnant for most if not all readers. But in the end, I am sure that at least a few, even most of you, will agree with me that lamentably there is no other option when it comes to the strange, invincible force(s) that "made" and control the inscrutable enigmas of what, where, and why we are. In this matter, the completely ignorant is as knowledgeable as the keenest intellectual because of the absolute impotence of both myth and truth, of fantasy and reality. Each of us is compelled to rely on a unique capacity for unbiased, honest reflection in coming to a satisfactory conclusion, a solution worthy of action untainted by the secular sin of lying to oneself. Below is an elaboration of what to me is the *only* incontestable solution to a problem of magnitude a million ("Mega") times more difficult than any other that has bedeviled humankind.

But are we up to it? Personally, I doubt it, but it's the only honorable action available to us in view of the daunting necessity. And even if it fails, the mere attempt may suffice to warrant dissuasion of posterity's chroniclers from changing our species name from Homo sapiens, man the sage, the knowing, the wise, to Homo sap, man the ignorant, the dolt, the dodo. In short, simply man the SAP he is now.

Chapter 5

Why the Mega-Problem?

I.

To humankind, every birth is an awe-full
"Priceless" treasure that to Nature is awful
Displeasure because
In obeying God's laws
That new human for Nature is offal.

II.

Our Unbridled Technology shows
How its power to spawn future woes
Can soon threaten existence
Through jobless subsistence,
Polluting as populace grows.

The solution sought by our over-hyped cognoscenti is really quite simple and has become "Mega" only because of modern society's short-sightedness, temerity, and demeaning arrogance. How could it be anything but simple in view of the very obvious fact that the problem to be solved is simple? Even Alley Oop, Oona, and their contemporaries probably knew it, and some of them actually applied the solution—for a while anyway. That solution was judged to be "sinful" by human's misuse, or really abuse, of his precious powers of reasoning coupled, of course, with the atrophy of his common sense that all animals possess instinctively.

Because of the human fear of death, our dread of helplessly succumbing to Nature's terrifying, often violent power, we naively directed our reasoning into the realm of bizarre but comforting mythology that we later labeled as "religion." The die had been cast to set the stage for our species' ultimate doom, essentially through suicide (as you will see later) by obeying the mandates of our ludicrously egotistical self-assumed "divinity." Armed with that "divine" sanction, the arrogant creature obediently and blithely goes forward to multiply our kind without restraint or qualms and to usurp at will all the biotic bounty of our self-proclaimed minion, Nature. Human reasoning power (?) thus crippled by the mythological "straightjacket" we chose to fashion for ourselves couldn't help but foster the enduring threat of the easily identifiable Mega-Problem and its corollary, the equally easily identifiable Mega-Solution.

Let's first take a look at some of the mechanisms operative in the Mega-Problem's development. According to Durant, history doesn't tell us just when human society passed from hunting to agriculture. [45] He surmised it to have been in the Neolithic Age, and that "the new regime...changed some old virtues into vices. Industriousness became more vital than bravery, regularity and thrift more profitable than war or violence. Children became economic assets, the more, the better. Early marriage and multiple maternity became desirable and the norm." No one had to beg for a "job" as is the case today. On the contrary, there was a labor shortage, which was a major factor in the institution of slavery.

---

[45] Durant, Will & Ariel: *The Lessons of History*. Simon & Schuster, NY, 1938, 38.

Opportunities for employment were abundant on seemingly endless tracts of land to be cultivated or deforested for farmland. And Jaspers pointed out that localized population pressures were easily relieved by migration and settlement into sparsely populated areas, "by colonization, or by conquest." Yet a much reduced amount of migration goes on even today, despite the much reduced amount of barely arable land still available, tragically resulting in the instability caused by marked overcrowding and localized chaos today.

The pace of the Mega-Problem's evolution was markedly accelerated by the advent of the much hailed Industrial Revolution, which, Durant said, "[O]nce again changed the economic form and moral superstructure of European and American life...Children were no longer assets and often left home; women were 'emancipated' on contraceptives; marriage was delayed (and often ignored) as the old agricultural moral code began to die. And wars added to moral laxity...The burgeoning inventiveness and entrepreneurship promised a bright future of a more comfortable and secure lifestyle capable of accommodating large populations. However, in this euphoric atmosphere, an important point was being overlooked by a purblind society, and that flaw was identified clearly and pointedly by economist Henry Hazlitt in his book published in 1962, *Economics in One Lesson*: "The art of economics consists in looking not merely at the immediate but at the larger effects of any act or policy; it consists in tracing the consequences of that policy not merely for one group but all groups...For depth in economics consists in looking for all the consequences of a policy instead of merely resting one's gaze on those immediately visible."[46]

---

[46] Hazlitt, Henry: *Economics in One Lesson*. Three Rivers Press, NY, 1979, 194.

A half century ago, already aware of Hazlitt's advice through simple reasoning, I was invited to address the luncheon meeting of one of those benevolent fraternal organizations. No specific topic had been suggested, but I assumed that they wished to hear about my work and perhaps research at a local hospital. It was a casual address that somehow drifted into an assessment of twentieth century life. As if instinctually aware of Hazlitt's insight, I decided to test it for the first time at that meeting, realizing that it most likely would be received by some of the audience with skepticism at best. I couched it metaphorically with this blunt statement: "In the future, the two most cursed names will be Louis Pasteur and Henry Ford." Surprisingly, the reaction was rather mutely and politely subdued with no opinions or discussion offered. Perhaps they were a bit shocked by my audacious, even "crazy" opinion, or they were reticent to expose themselves to the ridicule of agreeing with a highly unpopular notion that I am sure will be validated in the very near future.

Some historians credit the gold finds in Central and South America during the 16th century for having enhanced capitalism, already underway then, and eventually to the rise of the Industrial Revolution.[47] In anthropologist Hooton's emphatic opinion that merely worsened the problem: "Man's mechanical science has so far outstripped his biological status and his social ideals and behavior that it has become a menace...The mechanical contrivances which eliminate the necessity of thought, judgment, and skill in the user are almost as equally available for the use of the moron and the criminal as for the intelligent and the social-minded...The vast potentialities of these tools are

[47] Loewen, James W.: *Lies My Teacher Told Me*. W. W. Norton & Co., NY, 1995, 59.

used more and more for evil as they become perfected, cheaper, and more easily accessible. *Machines get better and better, while man gets worse and worse*"[48] (emphasis mine). Martin Luther King, Jr., expressed the same opinion with this short pithy statement: "Our scientific power has outrun our spiritual power. We have guided missiles and misguided men."

This is sickeningly obvious by the incessant increase in the quality as well as quantity of death, destruction, and all sorts of human misery fostered by high technology's incredible impact on the means of modern warfare. But perhaps the shortest, most incisive criticism of technology's almost universally lauded potential to augur only good for humankind came from the Polish poet, Stanislaus Lec: "Is it progress if a cannibal uses a knife and fork?"[49] Hooton pointed out further that "these mechanical marvels are the means of worldwide spread of vulgarity, depravity, and misinformation because they pander to the taste of those who are capable neither of contriving nor understanding, but only of crude sensory perception by turning a switch or pressing a button—motor performances within the capacity of a lemur."

He cited another factor as being of equal, if not more, importance in the evolution of our current and future Mega-Problem. Hooton expressed it best: "Medicine today is an extension of the maternal instinct mixed up with scientific techniques. It operates in an odor of sanctity and formaldehyde. Any objective criticism of medical science is difficult, merely because everyone of us is, or has been, or hopes to be, a grateful patient...The real question

---

[48] Hooton, Ernest A.: *Twilight of Man*. G. P. Putnam & Sons, NY, 1939, 302.

[49] Peter, Laurence J.: *Peter's Quotations*. William Morrow & Co., Inc., NY, 1977, 410.

at issue is not what advantage you and I individually may have enjoyed from medical science, but rather the value of the cumulative effect of medical ministrations upon the entire human species. Medicine has alleviated suffering and prolonged life; but it has, in so doing, also prolonged suffering and nullified the purging effect of natural selection. It has saved hundreds of thousands of debilitated organisms which are adding to the burden of society by reproducing more and more offspring...Medical science has committed this sin against the species because its members are not sufficiently enlightened or not courageous enough to declare their independence of the dogma of the sanctity of human life...They seem to have no clear conception of what they are doing to man, of what man is, or what he ought to be. They have hung back and let others lead whenever unpopular biological measures are necessary....I call upon this profession, which is actually directing the course of human evolution downward, *to reflect upon the wages of biological sin*" (emphasis mine).

Hooton would have relied on the application of knowledge gained from research in human heredity, which at that time was relatively rudimentary. Great strides have been made in genetic research in the subsequent six decades since then; and the Problem only got worse. In fact, it is now a full-blown worldwide crisis of wars, genocides, crime, and even incipient global anarchy. Hooton's dire warning—"The wages of biological sin is evolutionary death"—had better be heeded.

For most of my adult life, I have been aware, always subconsciously if not consciously, of all the things Hooton covered and expressed superbly. My last medical action, at the age of 72, brought me face to face with the shameless hypocrisy, venality, and dissembling on the part of the specious intelligentsia called rulers designated by their pathetically unconscious slaves called voters. That not only involuntarily ended my medical "transgressions" together with most of my inculcated illusions, but also it exposed me to an unexpectedly fortuitous experience. Having the advantage of much idle time for protracted and honest reflection during the meaningless and usually wasted 100 month interval of incarceration near the end of my life, and being, therefore, completely free with nothing but my life to lose, my thoughts gelled into a firm commitment on two issues: one important, the other of *extreme* importance.

Having limited resources, support, and circumstances, I decided to undertake a campaign of open and uncompromising advocacy of what has now become for me only a Mini-Problem with regard to the outrageously protracted suppression of our basic natural rights *explicitly* laid out by the Founding Fathers in the Ninth Amendment of the Constitution's Bill of Rights. Later, I further decided to openly advocate vehement denunciation of our thoroughly corrupt and omnipotent Supreme Court and equally openly to forcefully advocate diminution of its power, which Jefferson feared as being "despotic" and wished to reduce. He admitted that his failure to do so was one of his greatest regrets.[50]

I, too, expect to fail; but I will never stop pointing out the Court's appalling arrogance, dereliction of duty, and shameless dismissal of the highest law of the land in continuing to oppress the American people with its blasé and transparent neglect of its specific judical obligation. The unconscionable black-robed would-be tyrants refuse to invoke the hitherto *unused* Ninth Amendment in all cases concerning Constitutional rights. Instead, the shifty Court lamely relies for its rescue on the application of arbitrarily cobbled and sham "amendments," stressing the vague and harmful "due process" policy that any really honest Court would immediately and scornfully dismiss as

---

[50] Crawford, Alan Pell: *Twilight at Monticello*. Random House, NY, 2008, 4.

absolutely evil and unnecessary ploys. At bottom, this Mini-Problem is due to the Founders' failure in some way to limit the Court's undeserved omnipotence, perhaps a forgivably unintentional lapse of the Founders' usually clear foresight in an ambience conditioned by the euphoria of their newly acquired freedom. It's a fatal error that only a Constitutional amendment can correct. And the odds for that are slim and, indeed, out of sight.

The only other way to force the Court to invoke the Ninth Amendment is equally remote: to find a *nonviolent* way to acquire the requisite "standing" to bring the issue to Court by committing some kind of "capital" criminal act, such as breaking a law without physically hurting anyone or property. A charge of treason probably wouldn't do. And I'm sure even a *perfectly ideal* case would be shunned by the pusillanimous Court.

Chapter 6

The Mini-Solution (I)

In opting for Hoyle's clever route,
Man will doom himself, without a doubt,
To endure cycles rife
With their pain, woe, and strife
Till he dies in the last gruesome fallout.

The inimical effects of overpopulation are far more pervasive than most people realize. As human population numbers rise (see Fig. 1), it stands to reason that extra strain on the environment is to be expected. The exponential rise is apparent; the curve displayed in Fig. 1 shows a striking, even shocking, increasingly vertical course indicative of an acceleration in the rate of increase from which one can estimate that mathematician Hoyle's projected critical population level of 25 billion may be reached in about 150 years (barring an unforeseeable "Black Swan"). Also depicted is a similar but much slower accelerating rise in the number during the Middle Ages that was precipitously aborted by the Black Death (a classical "Black Swan") that dramatically (and brutally) reduced the population between the years 1347-1350 and ended about as suddenly as it started. Much more about that later.

"The key to extinction," says biologist Paul Ehrlich, "is change in either the physical or biological environment," and he blames humanity as the major agent of extinction: "Many of man's developmental projects yield short-term benefits for a few and pass on long-term costs to society. Species are disappearing (perhaps one thousand or more per year), much more rapidly than they're appearing....Tropical rain forests also are much more vulnerable to irreversible damage than forests in temperate zones."[51] Yet they are being recklessly exploited for farming, lumbering, ranching, and fire wood. Furthermore, laws and pro-environmental policies of the government (now euphemistically called "green") may have retarded environmental deterioration in the United States, but haven't stopped it. It is easy to overlook the damage being done in the oceans where many large fishes and sea mammals are being torturously killed by accidental entrapment in large nets being heedlessly used by commercial fishing firms, and sea turtles being horribly choked to death by ingested plastic bags discarded at sea.

Ehrlich says that the greatest threat to species and their environment is pork-barrel politics—the TVA Tellico Dam project in Tennessee for example. And the case of the tiny snail darter is "the first known deliberate act to push a species to extinction." Furthermore, the apparently unending debate over global control of so-called greenhouse gases drags on despite physicist John Holdren's ominous opinion that in just one decade, in 2020, "a CO_2 / climate-induced famine could kill as many as a billion people."[52] Finally, it has been claimed that "following a thermonuclear war....humanity would of necessity return to being a species of hunters, gatherers, and simple farmers." That may be a mistaken and short-sighted view because the massive radioactive fallout and terrestrial contamination would make such use of

---

[51] Ehrlich, Paul & Anne: *Extinction*. Random House, NY, 1981, 90, 159.

[52] Ibid., 90.

desolated land for habitation or farming dangerous or impossible perhaps for quite a long and unpredictable time. Ehrlich's more optimistic outlook thus seems to directly contradict Hoyle's contention that a better human being, a kind of new species, would emerge from his envisioned catastrophic setting, but now in contaminated environments rife with the likely, persistent, almost universal threat of agonizing death. We could expect each generation to experience severe physical deformity or congenital mental deficiency as a basic staple of reproduction.

Another horrible consequence of overpopulation is so-called crimes against humanity, which often culminate in outright genocide aimed at the elimination of an entire specific community, race, nationality, or sect. Just a few of the familiar and thorough attempts in recent times that resulted in the often brutal slaughter of millions include the World War One Armenian genocide perpetrated by the Ottoman Turks, the World War Two Jewish holocaust at the hands of German Nazis, the Rwandan tragedy in Africa aimed at "ethnic cleansing" of a minority group, the "killing fields" in Cambodia, and the slaughter in the Balkan civil wars. In each instance, the mass killing was followed by international exhortation to remember the massive crimes in an effort to prevent future repetition. The warnings never worked.

Human nature's darker impulses assure that warnings will not prevent genocide, and the recent, agonizing purges in the Darfur region of Africa is only the most recent example. Punishment of the responsible leaders was perfectly legal and entirely satisfactory only in the case of the Jewish tragedy; the victims and surviving relatives of victims were in part compensated by the nation of the convicted perpetrators who were duly punished. None of that happened in the other genocides, although several of the top Ottoman leaders were tried by a post-war Nuremberg type of a contrived, merely "eye-wash" tribunal. None of those responsible was convicted or punished, and the value of the victims' confiscated property was never replaced.

Only two of the other culpable leaders actually paid for their "crimes against humanity"; according to an article by Fred Edwords,[53] Jean Kambanda, the former prime minister of Rwanda, pleaded guilty for having committed genocide and other related crimes and was sentenced to life imprisonment; also, the Serb leader, Milosevic, faces trial for having committed the same crime in Kosovo. Other such tribunals in various countries, including the International Court of Justice in the Hague, have displayed the usual impotence to enforce court rulings.

Despite the recent indictment by The International Court of Justice of the leader of Sudan for conspiracy to commit a "crime against humanity" *still* going on in Darfur, I predict with the utmost confidence that nothing will come of it at least as far as the victims and their kin are concerned. Once again, the invincibility of the evil in human nature will prevail in this incredibly hypocritical world that will abide the outrageous contempt openly displayed by the Sudanese leader against the Court's apparently unenforceable indictments.

It has been suggested in an article by Hannah Hussey that the true roots of the Darfur debacle lie in the relationship between farmers, usually labeled as African, and nomadic herders described as Arab.[54] They have coexisted peacefully for years; however, some serious environmental factors caused this relationship to

---

[53] Edwords, Fred: *Why Genocide?. The Humanist*, Jan-Feb. 2008, 24-33.

[54] Hussey, Hannah: *No Water, No Peace. The Humanist*, Jan-Feb. 2008, 31-33.

disintegrate and allow political and ethnic tensions to explode into a real genocide. Severe drought developed in the latter part of the 20th century, forcing herders to travel farther south in search of water and grazing areas in Darfur. The growing human and animal populations, coupled with desertification made arable land a precious resource. The dry conditions destroyed the spirit of cooperation and interdependence. The increasing drought and desertification led to the next brutal stage of the conflict.

Over a million people are now living in ramshackle camps located at sites suitable only for sparse habitation and worsened by dwindling reserves of groundwater. The conflict has had an adverse effect on the environment and the animals, leading to increased poaching to near extinction in some areas. This is all taking place with the knowledge of the international community who, like observers of the prior genocides, has failed to take any real action to stop the suffering. But it should be pointed out that a humanitarian step to save lives will do nothing to eliminate the constant threat of even worse genocides in the future born of inevitable growth in ever higher population numbers of starving people.

I disagree with the opinion of noted anthropologist Margaret Mead who said, "I do not think that any government should—or can—legislate the number of children people may have."[55] On this matter, in this world, her opinion matters much less than the *force* of government that can do anything it wills, legal or illegal, moral or immoral. And George Washington is reported to have clearly and forcefully stated why it can: "Government

---

[55] Metraux, Rhonda: *Margaret Mead: Some Personal Views.* Walker & Co., NY. 1979, 107.

is not reason, it is not eloquence. It is force. And force, like fire, is a dangerous servant and a fearful master."

For many years, I have insisted that the day *inevitably* will come when society and parents will be *forced* to make a stupendous decision. For the sake of survival, all prospective and actual parents acting as proxy for their newborn or as yet unborn offspring will have to decide whether the latter will be part of a select, relatively small group of much greater, perhaps even unlimited, time span, to live as long a life as possible or, instead, be part of another select, probably much larger group of perhaps an unlimited number of individuals authorized to live only a much shorter life. All governments of the future *will thus involuntarily legislate* a suitable size of the population always kept constant and always in balance with Nature's biota. This *will* happen(!) unless humans desire to commit species suicide as a severely pauperized Nature struggles to survive with what's left of its decimated trove of innocent plants and animals, compelled to cope with a prior precarious, eons-long balanced existence in a now reeking global cesspool. Poor Nature—Poorer humanity.

Chapter 7

The Mega-Problem's Fallout

This problem's eventual fallout
Will mean death of man, too, without a doubt;
But the suicide way
To his fearful mind may
Make him choose Hoyle's plan as his route.

The probable results of massive overpopulation are manifold, and those who live long enough will most likely have the sobering opportunity to witness its horrific effect on the terrestrial existence of both humans and Nature. Because what is coming has *never* occurred before, as far as I know, in the human history of our present era, I must use my own knowledge, experience, and capacity for reasonable imagination to fantasize a coherent and plausible series of potential events as they might happen.

In the beginning, somewhere on the crowded globe will appear increasingly ominous signs of social unrest—like now occurring in Africa. From my first-hand experience, I have concluded that, in all likelihood, we have already reached that stage. During my futile, low-key, and audacious campaign for election to the House of Congress with the vain hope of instilling some honest ratiocination and common sense into what President Kennedy called a "zoo,"[56] I spoke to many potential voters. The vast majority seemed to be grateful for the opportunity to vent their seemingly helpless concern over a lingering, uneasy feeling that "something is wrong" with the condition of the economic and political life in our country, as well as in the world. I learned that the public is aware of something potentially dire now happening, and that their open expression of concern and frustration is tinged with tacit fear and even anger over their inability to identify what it is, what caused it, and why the apparently impotent authorities can't or won't talk honestly about it. After much introspection, I hit upon the likely answers to those questions, and they mandated widespread dissemination in a book like this hopefully to motivate merely free and open discussion and assessment of its critically important message.

That "uneasy feeling" is now being corroborated by numerous pockets of hunger and starvation, primarily in Africa where the process of decline has reached the level of outright anarchy—which is defined as the complete absence of government and law associated with political disorder and violence. Journalist Robert Kaplan, who has covered events in several continents, in his recently published book entitled *The Coming Anarchy*, paints a gloomy picture of what is in store for the world, heralded by what is already going on in the Middle East, in India, and especially in Africa.[57] The situation in the latter borders on outright chaos, bringing with it rampant unemployment and crime of all kinds, famines, incessant wars, widespread deforestation and desertification, water pollution and

---

[56] Carruth, Gorton: *The Harper Book of American Quotations*. Harper & Row, NY, 1988, 151.

[57] Kaplan, Robert D.: *The Coming Anarchy*. Vintage Books, NY, 2000, 4-12.

shortages, breakdown of unstable social and family structures, and of defenseless national boundaries ignored by undisciplined, widely migrating hordes—and throughout it all, and even worse, the wanton increase in what will end up being a massively starving population beyond help. And it will eventually be a global debacle of catastrophic proportions, involving the United States.

"The worsening African situation forebodes that something more terrible awaits us." That last point I am convinced is correct and justifies Kaplan's conclusion that *primitivized humans* will exist in societies at war due to resource scarcity and overcrowding: "The future map, the last one, will be an ever-mutating representation of chaos…The concept of a nation-state has been more fragile in the U. S. due to its multiethnic society; and therefore, it may not survive the next century in exactly its present form. Africa's distress will have a destabilizing effect on the U. S., resulting in insulated communities and cultures."

This supports my prediction made over a decade ago that in the very near future the U. S. may be broken into perhaps more than one other "nation-state" based on the ethnic density of the areas concerned. I also agree with his other conclusions that "the UN represents…the illusion of millions who want to escape from the historical cycle of war and power politics, and deny that the UN is a forum for the same power politics in disguise…and that recently expanded NATO is a variation of the Western Holy Roman Empire."[58] And we know what eventually happened to that political monstrosity.

When all is said and done, the fundamental cause of most, if not all, of society's wrenching crises and humankind's agonizing misery will turn out to be the invisible and easily overlooked phantasm of

---

[58] Ibid., 48-54.

overpopulation. Most of the public and even some economists are either not, or pretend not to be, aware of this decisive delusion. The people are kept ignorant of the cause of our current global economic "crash" with its massive unemployment and mortgage foreclosures. The cause and solution of the crash have supposedly "baffled" even the most respected private and governmental economists, and many are under the sham and vacuous hope that throwing trillions of newly "minted" artificial dollars at it will somehow by accident solve it; there exists a collective hope that by thus "doing something" as an "expert" to reassure the people will reduce public anxiety and fear. "Recessions" are really overpopulation; but its *real* solution is practically impossible to try to apply. This I call the Mega-Solution, which will be discussed in the next chapter. The implementation of the Mini-solution, which promises hardly any better chances of success in the long-run is at least acceptable and apparently "reasonable" enough to give it a try. But first, let's see how and why the Mega-Problem arose.

My father was an untutored Armenian subject of the Ottoman Empire who immigrated to America in 1912 with a few compatriots having no significant skills or knowledge of English. Yet all of them immediately and easily found employment in foundries, auto plants, or construction. The work was hard, to which they were accustomed, and the income was sufficient to establish and raise families. They didn't have to beg or whine for a job, and neither did the colonists when this country was born. In fact, there was a labor shortage, which led to the unfortunate and fundamentally unnecessary institution of slavery, designed chiefly for the profitable benefit of a pseudo-aristocratic elite composed primarily of well-educated European immigrants. As the population began to grow more rapidly by immigration of the lower classes, especially during "booms" like that of the "Roaring Twenties," employment was no problem as long as the labor force and the labor demand stayed in reasonable balance. It was upset by the drop in demand for labor that ushered in the Great Depression.

The question now is what caused the drop? In all likelihood, one of at least two major causative factors were and still are involved. One of them is manipulation by the invisible bureaucracy that controls and operates the economy and the money markets. We'll get to the second in a moment. Such manipulation is in direct contravention of Jefferson's warning when the country was established that, "We must not let our leaders load us with perpetual debt. We must make our election between economy and liberty, or profusion and servitude. Banking establishments are more dangerous than standing armies; and the principle of spending money to be paid by posterity, under the name of funding (today called 'bail-out'), is but swindling futurity on a large scale."[59]

It was the dangerous bankers who began and encouraged the wanton use of credit that could easily be manipulated. They offered credit without the restraint of responsibility and honesty. In the present financial crisis at least, such practices cost people their jobs. The government, in the transparent charade to show an ostensibly real intent to "right" the situation, used the suffering people's tax payments to reestablish liquidity of their banks and, probably once again corruptly, loaned money to reanimate cash flow—all in flagrant disregard of Jefferson's sage warning. Together with governmental collusion, banks have shackled future generations with the extremely burdensome onus of having to be heavily taxed to be able at least to try to repay that enormous debt

---

[59] Fuller, Edmund, Ed.: *Thesaurus of Quotations*. Crown Publishers, NY, 1941, 379.

dictatorially thrust upon a naïve, sheep-like society. And it is easy to guess who profits from all these sham gyrations.

It is tragic that the public is not aware of a startling fact, little if ever discussed—the fact that the staggering debt incurred by taxpayers to bail out the crooks in the fairly recent Savings and Loan fiasco is being paid off *yearly* by the equivalent of every penny of tax paid by every resident of Ohio, Vermont, Washington, South Carolina, and New Mexico who earns up to fifty thousand dollars per year—for the past 16 years and for the next 15 years until totally paid off in 2025![60] Now, on top of that, trillions of dollars of debt have been dumped on the groaning taxpayers; and despite the past wrenching experience, which obviously did not prevent the current, even worse economic crisis, the duped public was benumbed into sheepishly allowing the meretricious government more firmly to shackle them with the unbelievable debt of unimaginable amounts of money that, as hard experience shows, will inevitably bring about crippling monetary inflation that cannot help but accompany future, even more devastating, economic crises, the real, but supposedly "unprecedented" causes of which are plainly explicable.

It is obvious that our absolute "master" comprised of a self-serving, greedy, and unconscionably evil cabal organized by banking and industrial grandees plot and manipulate at will how and when the flow of the "life blood" of commerce, money, will "clot" to form the deadly "thrombus" and the complete paralysis (and amnesia) of a very effective societal "stroke"—indeed, such maneuverings offer a superb mechanism of control over their restless "herds."

---

[60] Bartlett, Donald L. & Steele, James B.: *America: What Went Wrong?* Andres & McMeel, K.C., 1993, 105.

In view of all this, the theoretical gist of the Mini-Solution becomes obvious, but I'm sure the reader will agree that its implementation is highly unlikely because of human nature; and we all have lived in this "culture," subjected to a lifetime of assault by manipulative propaganda exhorting us to buy and borrow to "stimulate the economy." This point didn't escape the attention of attorney Gerry Spence who summed it up nicely: "We are bombarded incessantly with a propaganda of things. So long as the media in the hands of the corporate master convinces us that we are free—mostly free to buy, to encumber ourselves to buy, to mortgage the homestead to buy; and so long as the corporate master's propaganda *spurs us on to buy and buy until we die*, we shall remain in the zoo."[61]

I have no doubt that advocacy of what will almost certainly be judged to be too outrageously contentious will lead to my vilification and to outright rejection of the idea without even superficial discussion. So, here briefly, is my conception of a Mini-Solution: enactment of laws or enforceable regulations to maintain a constant population density at a level commensurate with the concomitant sociological, political, economic, demographic, and biospherical conditions known to be capable of sustaining that level. A somewhat more than "Mini" solution would enable even better control in the enforced *reduction* of an established level if it is threatened by unforeseen circumstances, either human-made or naturally occurring. I realize that this suggested option may prove to be absolute anathema to an essential spirit of collective cooperation.

That calls for adoption of the Mega-Solution, the *only* one that offers any hope at all of absolute and complete

---

[61] Spence, Gerry: *Give Me Liberty!* St. Martin's Press, NY, 1998, 235.

success. I'm sure that many will deem it to be ridiculously Draconian. But the time will come, and fairly soon, when the neglect will be rued by some, a few of whom may decide to apply it personally for themselves. That will take extreme courage and firm commitment at best and perhaps cryptic insanity at worst. It doesn't matter; in time, the end result will be the same with either Solution, although one of the options entails the goodness of a bit of compassion for Nature and all of its biota: the only chance of allowing our species to survive for a while in a familiar environment kept at a suitable level of population having a reasonably sensible and tolerable lifestyle free of the inanity of excessive ostentation punctuated by spells of vacuous, wasteful, and crippling faddishness. It is the choice that a spoiled human nature will prefer, and the latter's penchant for flippancy will pose a constant threat to such a society's stability. But I'm sure that we will risk such a Mini-Solution just as we occasionally dared in the primitive immemorial past.

Chapter 8

The Mega-Solution (II)

Though a cyclic recurrence implies
That it's death through which new births arise,
Still it all comes to naught
Once extinction is wrought
For then all, even death itself, dies.

The above limerick is based on a very profound statement of fact by the noted author Jonathan Schell, which I discovered several years ago. It embodies an extremely impressive and extraordinary insight beautifully expressed: said Schell, "In one sense extinction is less terrible than death, since extinction can be avoided, while death is inevitable; but in another sense extinction is more terrible—is the more radical nothingness—because extinction ends death just as it ends birth and life. Death is only death; extinction is the death of death."

Not many of us know what chaos really implies: it is usually defined as confusion, a mixed mass without due form or order. Life under such conditions is unimaginable if not experienced first-hand. And it is inevitably coming for all humanity to experience as part of a cataclysm of monumental magnitude caused by the "crime" of widespread "biocide" and the involuntary suicide now being senselessly committed by the kind of serial biological killers called us. We use, as if a toy, our ill-conceived and uncontrollable super weapon known as *massive* overpopulation. Its appearance on earth has already occurred, primarily among the poverty-stricken in several densely peopled countries and especially in Africa. Kaplan has documented in great detail the increasingly ominous African situation that heralds what Kaplan says is in store quite soon for the entire world, including the affluent countries.

In Kaplan's well documented opinion, crime in West Africa "makes a natural point of departure with regard to what the political character is likely to be in the 21st century.... an eerie taste of what American cities might be like in the future, with hordes of similar young men everywhere in a very unstable social environment clearly on the verge of igniting." He blamed Africa's high birth rate, with its resultant deforestation and desertification, on loose family structures and on the explosive spread of the HIV virus. The ecological damage in turn led to soil erosion, flooding, and an increase in the prevalence of rapidly resurgent antibiotic-resistant diseases such as malaria and tuberculosis. "The streets of expanding cities and their ballooning populations are long puddles of floating garbage," and their populations are doubling in a little more than a generation. Increasing environmental scarcity of cropland, fish, forests, and water has led to prolonged and brutal ethnic, racial, and cultural (often localized protean guerilla-type) wars and atrocities of wanton rape and mayhem, including mutilation of kidnapped children forced to fight as ordinary soldiers. Displaced refugees form wandering groups that ignore geographic national borders, resulting in purely *de facto* borders that are delineated by the hordes, despite the almost nonexistent controlling authority of weak and corrupt governments.[62]

---

[62] Kaplan, Robert D.: *The Coming Anarchy*. Vintage Books, NY, 2000, 26.

This is a preview of the kind of chaos the skyrocketing populations of initially so-called underdeveloped nations are beginning to experience and eventually all supposedly civilized nations of the entire globe will get to know first-hand. The ultimate scenario will be outright anarchy with its unspeakable horrors. Those who are familiar with details of the so-called Black Death of 1347-1351 will have a fairly accurate idea of what to expect in the coming cataclysm, but probably take little or no account with respect to some special horrors that the Mega-Solution's appearance will bring. Although the cause of chaos and anarchy will be different in both cataclysms, the significance for, and impact on, human cultural evolution will be *radically* different.

The Black Death originated in the steppes of Mongolia and was transmitted very rapidly along well-traveled trade routes westward through the Middle East to Western Europe. It was caused by a then unknown lethal contagious disease caused by a hitherto undiscovered bacterium now named Yersinia pestis (formerly Pasteurella pestis). In 1347, that bacterium was carried on land to the West by traders whose merchandise inadvertently carried rats infested with the fleas that are the vector for the disease; they were carried by sea in rat-infested merchant ships from the Orient. It could also easily be spread by coughing and from contact with oozing blood from skin lesions caused by the disease.

Initial symptoms are flu-like with fever and chills for a week or two accompanied by swelling of lymph nodes called "buboes" (thus, "bubonic" plague) in the neck, armpits and groin. Bleeding welts that turn black appear on the skin, giving the disease its vernacular name of Black Death, spread of which was surprisingly rapid. Involvement of the lungs, the "pneumonic" form of the disease, resulted in extremely rapid and excruciating death. The plague spread with incredible speed and wide prevalence, eventually conquering England (where it ended the Feudal System) and Russia; the disease wiped out about half of the affected victims in densely populated European cities where death rates of up to 7,000 per day were common.

The unknown cause understandably terrified whole continents; even Pope Clement VI was threatened but refused to heed the advice of his personal physician to flee to a safer city or forest in the countryside that happened to be free of rats. He remained in his castle, surrounded by the flames and smoke from numerous large candles recommended by his physician, in the futile hope of "doing something," of cleansing the air of an imagined "miasma" believed at that time to be the bearer of deadly contagions. The fear of acquiring the disease through direct contact forced unaffected people to stop burying the many malodorous piles of decaying, bloody corpses. The healthy mountain and forest dwellers then found very lucrative jobs as the only willing grave diggers to handle interment in often mass graves without caskets, which rapidly became scarce.

The terrified people soon became hysterical, fighting, praying, raping, and even killing each other. Many chose to remain barricaded in their homes; and priests and physicians often refused to enter homes for fear of "catching" the disease; the clergy's reluctance encouraged many to lose faith in their religion. Abandoned homes were ransacked. Moral laxity fostered crime and debauchery of all kinds. The loss of faith for many was in some measure partly replaced by spontaneously large rowdy bands of so-called flagellants who marched through the land in irregular groups and columns of men, bare to the waist, beating their bloody backs and arms with spiked sticks and leather thongs, hoping to gain the mercy of God by emulating Christ's travail carrying the cross. They became very popular, the "people's rock stars," who hysterically broke into sealed churches, brutally attacked and even murdered priests, and assumed their role in mock and crude outdoor religious ceremonies.

Hatred of the scapegoat Jews was fanned by the belief that they were responsible for the catastrophe by poisoning the water, and they were mercilessly beaten and killed. The flagellants' growing power alarmed the Pope who then had them eliminated by the authorities. Even the Pope's famous physician, Guy de Chauliac, was infected; and he devised various ultimately ineffective means to cure himself while ministering to the people as best he could, always keeping meticulous notes like a seasoned researcher. The markedly reduced population (by one half in Europe) offered less opportunity for spread of the germ, and then the dreaded disease slowly died out in about 1351. The physician survived and joined other survivors in an astonishingly rapid and smooth resumption of their prior lifestyles.

The disease still exists around the world, even in the southwest U. S. where small rodents are plentiful; and localized outbreaks are held in check today by effectual antibiotic therapy and preventive measures.

The chaos of the coming anarchy will be significantly different. In the case of overpopulation, there would be no cause for the agonizing signs and symptoms of an acutely lethal infection. Instead, its untold misery will be due only to one serious factor: the painful lack of food and the absence of much opportunity to grow or purchase it. In such circumstances, it would be utterly impossible to defend farmland or carry on orderly commerce. Angry, hungry, and starving, perhaps heavily armed and determined marauding mobs, half-crazed in an ambience of lawless anarchy, would become highly dangerous predators driven instinctively and without much moral restraint to any action, usually violent, whether or not success were possible. With starving millions so engaged, the full impact of widespread depravity wrought by a now totally uninhibited and evil aspect of a normally reined human nature (as described above by Schopenhauer) will be painfully and starkly exhibited in an ugly panorama that will overwhelm the imagination.

All land and sea animals, birds, worms, insects, fruit-bearing trees and bushes will become scarce or disappear altogether. Zoos and wildlife reserves will be emptied almost immediately. Then the fury of the hunt will turn to easier prey through the means of murdering people—if necessary—to purposely contrived anthropophagy (eating humans, cannibalism) and necrophagy (eating corpses) and infantophagy as well as neonatophagy (eating babies and newborns), an act which Sir Francis Bacon is said to have advised, perhaps whimsically, as being beneficial in a time of famine. The nomenclature has been further refined; endocannibalism is defined as eating a member of one's own group; exocannibalism, the eating of an outsider or alien; and autocannibalism, eating one's own flesh (I wonder if this includes lip biters).

Anthropologist W. Arens advises caution when it comes to the many reports of the practice of cannibalism throughout history. Says he, "As a result of directed research, conversations with colleagues, and some deliberation, as an accepted practice for any time or place I am dubious of the actual existence of this grotesque act...The possibility that early Christians may have been cannibals is assumed to be false, never seriously considered. Recourse to cannibalism under survival conditions, or as a rare instance of antisocial behavior is not denied for any culture. But whenever it occurs it is considered a regrettable act rather than custom."[63]

---

[63] Arens, W.: *The Man-Eating Myth*. Oxford Univ. Press, NY, 1979, 9.

However, other social scientists reported that it *was* customary among many ancient tribes for several purposes, such as eating kinfolk as a sign of respect, to gain the enemy's strength, to avoid wasting life-saving "food," to execute a criminal ("judicial cannibalism") by certainly annihilating him, to enable post-partum women to regain their expended energy, or to merely conform to the accepted mores of the group. In fact, some tribes have so firm a cultural tradition of consuming their dead that they are just as surprised on learning that we put the corpses in the ground as we are surprised on learning that they eat them.

I predict that a new and extraordinary source of sustenance, obtained from the living human body, will appear during the coming overpopulation debacle. It will become commonplace and will be called "placentophagy," eating the afterbirth. I have observed at close range a cat that, immediately after having given birth, began eating part of the placenta and the attached cord. The instinct probably evolved by nature more for the purpose of establishing a firmer bond between parent and offspring rather than as a means to assuage hunger or to replenish the mother's nutritional sacrifice through pregnancy. If humans have or even had such a powerful instinct, humankind's long and persistent disuse, oppression, or emotional and sectarian bias may have obliterated any trace of it in the human genome. If only atrophied because of our current sectarian tyranny in this time of so-called humanitarianism, then the dormant instinct in humans should be resuscitated and nurtured as potential salvation of some human lives in occasionally rare but suitable exigent situations, like the one that now looms.

I seriously advise that this act be proclaimed through the legislature as a perfectly moral natural right (which it is according to our currently taboo Ninth Constitutional Amendment and especially according to

Nature's supposedly infallible "laws"), not only out of necessity during severe famines, but also as a routine culinary tradition if so desired. It's entirely a personal matter concerning a harmless, mindlessly discarded, and completely natural "manna" benevolently bestowed to animals by Nature. Then why not to the supremely bestial animal called human?

My occupation as a pathologist obligated me to closely examine, grossly and under the microscope, intact placentas with attached umbilical cords, routinely considered to be surgical specimens and therefore subject to mandated examination and eventual disposal. While describing to a stenographer the gross external and internal features of a placenta in one case, in which it looked so delicious, exactly like a thick, large, round, red, fat-free steak, that I was compelled to utter, quite seriously, a statement that made the secretary taking the dictation wince and reply with a mild objection that my indiscretion of having described it to be worthy of food for human consumption had made her feel a bit sickened. Nevertheless, placentophagy will be globally practiced during the coming riotous anarchy. And, therefore, unluckily, pregnant women will become the especially coveted targets for murder. It should be noted that some placentas today are not discarded as garbage and are used instead to prepare clotting factors for the treatment of certain blood disorders as well as for the procurement of stem cells.

Despite the repugnance of merely the thought of eating humans, most rational people would agree that there are unconditionally acceptable, even laudable, occasions when the desperate act must necessarily be performed. It is claimed that there developed the strange practice of what is called "medicinal cannibalism," consisting of bringing drinking bowls to occasions of beheadings to catch the human blood believed to be capable of reducing seizures and curing many illnesses. The controversial practice was condoned all over Europe during the time of the infamous Inquisition. It is even alleged that during the same period cannibalism was practiced in Ireland in the late 16th and early 17th centuries because of food shortages and famine. For the same reason, in 1594, in Paris, bread was said to have been made from human bones.

At about the same time, the famed controversial physician, Paracelsus, was prescribing "maiden's zenish" (menstrual blood) for the treatment of gout. Concurrently in Egypt, so-called "mummy" consisting of small, very expensive packets of mummified flesh were being touted for the enhancement of health; and it became all the rage in Europe. So-called "mummy" was prescribed by the best doctors whose faith in its apparent effectiveness was based on their recognition that most of the "cured" illnesses were not lethal and actually cured themselves, an age-old ploy that Voltaire well characterized: "The art of medicine consists of amusing the patient while nature cures the disease." Finally, today, human body parts and fluids are available in many of China's pharmacies.[64]

When it comes to a consideration of human consumption of the human body itself, very few individuals know that, according to a respected historian, it was practiced by early starving settlers in Virginia. They were bickering and lazy and spent their time digging for gold instead of planting and growing crops. The ensuing hunger drove them to dig up and rob Indian graves and to eat the corpses—a tasty mash of undoubtedly rotting flesh.[65] Many

---

[64] Travis-Henikoff, Carole A.: *Dinner with a Cannibal*. Santa Monica Press LLC, CA, 2008, 146.

[65] Loewen, James W.: *Lies My Teacher Told Me*. W. W. Norton & Co., NY, 1995, 90.

Indian tribes of North America were already practicing cannibalism when settlers landed on the eastern shores. Most evidence for this is found in the Four-Corners area in the West; there cannibalism was probably a result of drought and starvation.

Today, most people will recall having heard of the often-quoted legendary example involving three or more shipwrecked sailors on a life raft floating for days in the ocean. They ran out of food and were facing starvation when one of them died, and the dreaded question came up for discussion: "bury" the body at sea or consume it in an effort to survive—of course with the unspoken assumption that it would have been approved by the victim as God-given "manna" to his comrades. As mentioned above, Aren's warning should cause us to think twice before accepting the veracity of such plausible and compelling stories.

On the other hand, most people are equally unaware of credible accounts of cannibalism in our own time, verified through documented DNA studies. In the Pacific Islands during World War Two, the Japanese cannibalized downed U. S. pilots as well as their own dead comrades, a secret kept for 50 years. After the war, five high-ranking Japanese military personnel were executed secretly for having cannibalized U. S. servicemen.[66]

Nevertheless, there have been *at least three* fairly recent, sensational, instructive, and unequivocal instances of *individual* and *group* cannibalism, *well documented* in minute detail in separate books specifically about each case. The individual case might qualify as a special subgroup of the antisocial type of cannibalism mentioned by Arens, but instead, it was designated as "love

---

[66] Stampf, Guenter: *Interview with a Cannibal*. Phoenix Books, Inc., Beverly Hills, CA, 2008, 48-55.

cannibalism," which is briefly discussed below. And I was surprised to learn that a *third* very recent and well-documented case occurred in Austria in August 2007.

The well-known group event was an example of justifiable survival cannibalism. It happened on October 12, 1972. An airliner flying from Montevideo, Uruguay, to Santiago, Chile, encountered bad weather, and a sudden air pocket caused the plane to crash on a snowy Andes mountain top, killing 13 outright. Two of the wounded died the first night, two more in a few days. About two weeks later, an avalanche killed nine more, and 17 died seventy days after that. Starvation loomed as meager food supplies dwindled and hunger set in.

Then a second year medical student stepped forward and encouraged them to try to survive. On the tenth day following the crash, the majority of survivors for the first time tasted, and, with great trepidation, consumed human flesh. They had agreed on specific rules: no eating of one's friends or relatives, cut only small pieces, and no pilfering. The flesh was defrosted and partially cooked on the hot, sun-bathed fuselage of the plane. Many disliked fat, but they had to eat some daily to avoid constipation and to aid digestion.

At first, only muscle and fat were eaten; later, they supplemented their diet with hearts, livers, kidneys, and intestines (small intestines had a salty taste); but lungs, skin, brains, and genitals were shunned. In time, the compassion for the injured turned into loathing as they began to be looked upon as mere parasites. When meat ran out on the fiftieth day, they began scavenging previously rejected hands, feet, lungs, brains, and tongues—and in one case, the testicles. Even parts beginning to rot were now acceptable and tasted like cheese. Brains, which turned out to be the most nutritious, could not be eaten when rotting, so the heads were buried.

The survivors were ultimately rescued and having lost an average of fifty pounds each returned home more religiously philosophical and much less materialistic.[67] This is a stark example of what will probably become common during the Mega-Solution, together with other horrors which most people are not accustomed to contemplating, let alone witnessing—and, God forbid, actually acting out eagerly in grotesque detail.

The solitary episode, considered by some to be a case of "love cannibalism," involved a very recent, exceptionally bizarre combination of homosexuality and assisted suicide performed through voluntary cannibalism at the request, and with the voluntary participation of, the victim in his own home. Both were homosexuals who met through advertisements on the Internet. The cannibal, a 41-year-old German man, now is imprisoned and has been dubbed the "monster of Rottenberg." His victim, more accurately a partner, was a 36-year-old German engineer who wished to be killed and eaten and who enthusiastically participated in the bizarre repast until his throat was cut. The grotesque sacrifice took place in the spring of 2000. It's occurrence was discovered and reported two years later when the police located the cannibal through his advertisement for another man who also wished to die and be eaten.

The Austrian case involved a nineteen-year-old German who killed and ate parts of a homeless man, aged forty-nine, in a shelter in Vienna. The victim might have been killed two days before discovery by a cleaning woman. Police found and arrested the killer in the same blood-splattered room next to the disemboweled corpse with blood dripping from his mouth. The body had been

---

[67] Travis-Henikoff, Carole A.: *Dinner with a Cannibal*. Phoenix Books, Inc., Beverly Hills, CA, 2008, 48-55.

slit in many places, and the skull had been cracked open to allow access to parts of the brain. The killer had arranged the victim's organs on a plate in the adjoining room.[68]

These are shocking examples of the repugnant crimes and atrocities that will always characterize times of complete and widespread anarchy and starvation due to unavoidable chaos that massive, unstoppable over-population portends. The terrified, frustrated, angry, and extremely desperate populace will become inured to, and eagerly participate in, the most outrageous acts of infanticide, voluntary and involuntary suicide and euthanasia, and wanton rape, murder, and cannibalism—all of it on a grand, unprecedented scale.

The Mega-Solution will force emergency enactment of harshly punitive laws made absolutely necessary to hinder or prevent any possibility of survival. That will assure the disappearance of *all* health-related professionals and activities, of *all* research in every discipline, and of *all* formerly usual educational facilities and activities. In short, civilization as we know it will *shut down completely*. Only the procurement of universally scare food will remain paramount, of any importance or value; the exclusive aim in life for everybody of every class will be degraded; the good people of the world will become desperate, desultory, rapacious, lawless, and half-crazed nomadic mobs.

If humans are to survive on earth as a unique species, we will have to hope for, and somehow struggle to reify, the implicit success for our limited survival only so long as Hoyle's cyclical scheme remains effectual. But in the long run, the odds are against it because of the irony in this battle of survival with Nature: it is man's adversary,

---

[68] Stampf, Guenter: *Interview with a Cannibal*. Phoenix Books, Inc., Beverly Hills, CA, 2000, 130, 133.

"Mother" Nature herself, who has instilled in him the "time bomb" of his ultimately certain defeat, no matter which scheme he chooses to follow. That lethal "bomb" was identified by Hoyle: It's man's excessively eager capacity (and I would add motivation) for reproduction that is mediated through the irresistibly pleasurable means of coition impervious to any human-made contrivance to curtail or nullify it. The unlucky "tattered survivors" will have endured an experience that will have indelibly inculcated into humanity's collective mind or into humanity's deficient genome what all other creatures already know since birth—that life and its meretricious world just ain't beautiful.

Let's see how humans and Nature will fare in their epic battle if we choose to rely on Hoyle's scheme, which I call the Mini-Solution instead of relying on the only other available option, the Mega-Solution. In the end, both are rife with misery, but only the "Mini" has a highly desirable redeeming value for the human population; the other one, the "Mega," for Nature—all of which is elaborated in the next chapter.

On completion of each cycle, humanity's genetic health is expected to improve while progressively debilitating Nature through our mindlessly unending assault, resulting finally in much environmental damage, pollution, depletion of Nature's sometimes irreplaceable resources, and incredibly accelerated extinction of plant and animal species. Obviously, losses for Nature automatically mean some degree of loss for humans, too. Whether or not that loss is recompensed by human gain of a new species status through the expected improvement of intelligence and sociability is highly problematical. The cumulative effect of such losses through several cycles will magnify them faster for humans than for Nature and prove to be relatively more harmful for humans than for Nature, simply because, as mentioned earlier, human losses would be rather abstract and less capable of being replenished, if at all, whereas Nature's are concrete and have been able to be replenished since earth was formed. Furthermore, humans as a greatly improved new kind of species could hardly build a comparably suitable civilization with only the scanty or nil means of a now weaker or poorer Nature incapable of providing them. And besides, the constant battle will still continue to exacerbate humanity's disadvantageous position.

Thus, given enough time, as Nature weakens, humans will die. Our dogged, gradually less powerful persistence to maintain existence of our self-vaunted species through the gradual impoverishment via multiple cycles will thus have proved to be the lit fuse that will explode the "time bomb" of our suicide.

Now let's take a look at what the Mega-Solution promises and most likely will deliver.

Chapter 9

Fallout of the Mega-Solution

Earth's wounded NATURE can revivify
Its original pristine land, water, and sky;
And, too, its biota
Man made an iota,
But only if man has the good will to die.

"If under certain circumstances it is the duty of an individual to sacrifice his life for something higher than his life, might it not be the obligation of mankind (collectively) under certain circumstances to do the same?" The eminent psychiatrist-philosopher, Karl Jaspers, thought so and was among the few who had the courage to state such a view outright. He and a few others claim it to be "a sacrifice to futurity." The noted author, Jonathan Schell, tacitly agreed with and quoted Jaspers by asking: "Is it possible that one day it might be our obligation to set both human interests and human morality aside and destroy the species for the sake of God or Nature?" He, Jaspers, and a few others claim it to be a justified and laudable "sacrifice to futurity".[69]

We tend to oversimplify the real motive for suicide through categorization, which encourages a disregard of the motive's essentially un-dissectible complex of elements

---

[69] Schell, Jonathan: *The Fate of the Earth*. Stanford Univ. Press, Stanford, CA, 2000, 130, 133.

of various magnitudes and degrees of influence that determine the "firm foundation of a life worth living."[70]

Many people feel as did Tolstoy, who agreed with Schopenhauer that life is a "meaningless absurdity." And so did the noted screen actor George Sanders as he expressed in a suicide note: "Dear World, I am leaving because I am bored. I feel I have lived long enough. I am leaving you with your worries in this sweet cesspool. Good luck."[71] In addition to boredom, he was also afflicted with illness and financial problems. Perhaps somewhere in his motive was a prior faint awareness of the enigmatic idea of a sacrifice for the sake of eternity, for the sake of Nature's future. It might have instilled in him and others the exhilaration needed to nullify the absurdity and instead to replace it with the incontestable nobility of a sacrifice for the eternity of indispensable Nature—to give it the time necessary for its orderly self-replenishment of resources and for the restoration of its originally balanced integration of its "laws" and ecological constituents with the established existence of its surviving panoply of plants and animals. We may never know for sure.

But beyond such speculation over the *concern for the future of Nature*, it cannot be doubted that Sanders' suicide also is an example of a mixture of several motives: to escape what he deemed to be the vacuous *absurdity* of life; out of *necessity* to escape the intolerable suffering due to incurable illness; and of course the *desperation* of hopelessness—certainly not merely a cowardly act.

---

[70] Jaspers, Karl: *The Future of Mankind*. Univ. of Chicago Press, 1961, 172, 173.

[71] Van Der Beets, Richard: *George Sanders: An Exhausted Life*. Madison Books, NY, 1990, 203.

Another credible and striking example of such a drastic *individual* sacrifice is familiar to most of us already: the ritual, slow, and very painful, uniquely Japanese act of suicide called "seppuku" ("self-disembowelment"), better known outside of Japan as "hara-kiri" ("belly-cutting").[72] It evolved in the 12th century as part of the Samurai Warrior Code of Honor for those who, defeated in battle, wanted to avoid the dishonor of becoming a prisoner of war; sometimes, seppuku was a way to express the shame in having failed in doing duty, a means of protest, or a way to demonstrate loyalty. It manifests extraordinary self-discipline for the sake of futurity of personal and societal integrity of character, quite deficient in most Western people.

The ordinary act involves an individual using a short sword to stab himself in the left abdomen, cutting across to the right side, then upward. Its exemplary form required another self-stabbing in the lower chest to cut downward across the first horizontal cut and concluded, finally, with the stabbing of one's throat. As a form of judicial execution (abolished in 1873), it involved a friend or relative standing behind the prisoner and bearing a sword to act as a "second" to decollate the prisoner immediately when he reached for the sword to stab himself in the throat, thereby enabling the condemned to avoid the dishonor of having to submit to a common executioner. Seppuku was widely practiced by Japanese military personnel during World War Two and even, rarely, among a few civilian intellectuals after the war.

An even more striking and famous example of that kind of *collective* sacrifice occurred two millennia ago in the first century A. D. in a Roman fortress atop a hill called

---

[72] *Encyclopedia Britannica, 15th ed.* Vol. 10, 2005, 641.

Masada[73] where almost one thousand Jewish Zealots were hopelessly besieged and faced capture and enslavement by the Roman legions. Instead, they chose death. Using swords and knives, fathers "gave freedom" to their children, husbands to their wives, bachelors to fathers and husbands, and their ten leaders to the bachelors. The last Zealot tried to kill himself by falling on his shortened javelin. However, the shaft broke under his weight, leaving him propped up on his knees, bleeding profusely. The shaft then slipped on the blood, and he fell flat, screaming in pain, pleading for one of the five women, designated to hide and survive, to get a knife and end his misery. But he died anyway, whispering, "A last command—kill me! The pain gets between me and God. Kill me. Life is a...calamity." Could there be a more moving and eloquent example of the kind of sacrifice Jaspers had in mind, despite the unavoidably grotesque means of assisted suicide *and* euthanasia to achieve it?

Jaspers elaborated further that "man is born to be free, and the free life that he tries to save by all possible means is more than mere life. Hence, life in the sense of existence—individual life as well as all life—can be staked and sacrificed for the sake of a life that is worth living." If now, after all the lives that have been sacrificed in history, the total sacrifice of humankind appears possible, there remains one ambiguity: would it be an act of despair? *Despair* turns the man who is ready to die *daring,* in quest of a better life, into one who seeks death because he is tired of living. The philanthropic will to dare turns into a misanthropic will to destroy. Or would it be an act of *necessity,* arising from the unfathomable source of all things? It could be necessary only as a sacrifice made for

---

[73] Koseff, D.: *The Voices of Masada*. St. Martin's Press, NY, 1973, 231-237.

the sake of eternity. This sacrifice alone would be the firm foundation of a life worth living.

It seems as though "intelligent" humans either cannot, or simply choose not, to see that we cannot and will not "win" this cosmic battle, this rapine mandated by our foolish obedience to the outlandish mandates of our own concocted mythology—to "go forth and *multiply*" and to *exploit and selfishly usurp* all that the earth and its immortal Nature have placed at our specific disposal. And yet we sense the evil of such action and of its implied result: we have already devastated the earth, and we are killing it ("terracide").

Human vanity has driven us to outrageously wasteful research in space with the foolish hope of escaping the hell on earth we know we are creating. Jaspers has very clearly described this major foible of humans: "Can we have confidence in an escape from the technological menace through the technology of space? No, such ideas belong to the hubris of technological omnipotence...The idea of a new Noah's Ark could drastically limit survival through factors such as biological and environmental necessity," of being compelled to exist for eons in the confines of some sort of spaceship coursing through the limitless expanse of a universe in a blind hope of finding some kind of physical entity at least as generous and accessible as is our earth. Finally, there is "man's delusion of less than total doom on earth due to perpetual peace guaranteed by the absolute, and for him inconceivable, renunciation of warfare."[74]

But history and human nature have proved that that last hope, too, is futile. And besides, how could the killer, humans, live in the ashes of the world we killed—and why

---

[74] Jaspers, Karl: *The Future of Mankind*. Univ. of Chicago Press, 1961, 320, 323.

would we want to? I have long believed that the current extremely costly probing of space, triggered by the fantastic eidolum of physicist and pioneer of space research, Robert Goddard, is planned and guided by the same invisible, elite power that controls "all industry and commerce," and, indeed, the ultimate destiny of our world—in vain hope, explained by Jaspers, of avoiding human extinction by this self-appointed, select, unknown, *elite cabal* to perpetuate survival of H. sapiens through colonization of some other cosmic entity in the Solar System or somewhere else in the universe—vain because Havelock Ellis, who evidently understood the essence of human nature, had already mocked us for such folly: "The sun, the moon, and the stars would have disappeared long ago—as even their infinitely numerous analogues on the earth beneath are likely to disappear—had they happened to be within the reach of predatory human hands."[75] Such cosmic catastrophes would be inevitable if the existence of the human species and its unchanged nature were somehow to survive this first "gauntlet" of Hoyle's cycles on earth, but I consider that to be highly unlikely.

Fortunately, humans are mortal, but unfortunately the inherent evil of our nature is not. It dooms us either to the degrading fate of ignoble involuntary suicide forced upon us by our own wanton and lethal impoverishment of our means of sustenance in our hitherto nourishing environment; or to incomparably nobler *voluntary* suicide, on an individual as well as collective basis, as we sacrifice to futurity in the form of a legacy of inestimable benevolence to what certainly will be a very grateful Nature.

---

[75] Bartlett, John: *Familiar Quotations*. Little, Brown & Co., Boston, 1955, 784.

As Jaspers wisely concluded, "Collective suicide is not out of the question if leaders meet in loathing, or indifference, or in blind malicious and reckless destructiveness." A perfect example of the latter is that of Hitler who, seeing he was doomed, plainly wanted to doom the excessively obedient German people, too, [76] by "ordering" them, while he had the power, to continue utterly futile hostilities in an obviously lost cause and thereby commit collective suicide along with him. Fortunately, the war ended too suddenly for his diabolical aim to be promulgated.

I lived at that time as a 17-year-old awaiting a call from the Army draft, and I remember his words clearly: "The German people are not worthy of greatness such as mine!" No, a sacrifice to futurity must be first an individual decision, one might say as a sort of "romantic suicide" carried out for the love of Nature. Once that is achieved as a paradigm, it could be the "trigger" needed to spark enough motivation for a sincere, benevolent, collective species response.

In any case, the Mega-Problem, of course, doesn't involve loathing, malice, or destructiveness; but for the love of Nature, humans *must* vanish—and soon! Period! We simply gotta go!

---

[76] Jaspers, Karl: *The Future of Mankind*. Univ. of Chicago Press, 1961, 323.

Chapter 10

Non-Being

I.

That death which the faithful esteem
As a call from their Being Supreme
Is instead to a skeptic
Just plain narcoleptic
Deep sleep—with decay—and no dream.

II.

For life after death is a feint
Used by man to fend Nature's constraint,
Since whenever you doze
Without dreaming it shows
Just how *nothing* can be if *you ain't.*

As mentioned earlier, Jonathan Schell very clearly, concisely, and precisely differentiated the significance of non-being into two forms: death, that *we think* we know well from having caused so much of it on earth (but tells us nothing of its basic essence), and extinction, which is even a more abstract and inscrutable concept. To these enduring, probably insoluble, and certainly scientifically un-investigable enigmas, I would add a third: dreamless sleep, about which we likewise know absolutely nothing, *per se*, but only the occasionally nightly experiencing of a live consciousness when connected to a realistic dream or a weird nightmare. We are left completely ignorant without an inkling, with regard to the *same* essence, fundamental nature, or ultimate meaning of the *same* consequence of these three *different* mechanisms of non-being, constituting a weird conundrum, probably inaccessible forever to the vaunted probing of science.

So, what little do we know? For me, Jonathan Schell's analysis of the significance of extinction relative to human existence is the most cogent and compelling. Says he: "How should we gauge our worth as a species? For example, we cannot say in any simple or unqualified sense that the end of the world is bad and its continuation good, because mankind is not in itself good or evil but is the source of both. No stone, tree, or lion did anything ever good or evil. Neither can we say that mankind is useful—to or for whom or what?"

Absolute refusal to participate in evil, as Socrates did, might have been in the back of Jaspers' mind when he said that to avoid certain evils it is permissible to perpetrate extinction: *"The fact that it is not extinction but life that brings suffering and even death, is the clearest proof that extinction is misconceived as a disaster*" (emphasis mine). So said Jaspers: "On the contrary, *survival* means disaster, endlessly as long as life is beset by accident and folly…Respect for the earth is the foundation of all life. There is no 'sovereign' right to destroy the earthly creation on which all life depends for survival."

Jaspers continued with his extraordinary penetrating insight: "Self-extinction of our species is an act that, without quite admitting it to ourselves, we plan to commit in certain circumstances." He said that "being impossible as a fully intentional act, it can come about only as a result of inadvertence, as a 'side effect' of some action that we do intend, such as the defense of our nation, or liberty, of socialism, or whatever we happen to believe in. Our failure to acknowledge the magnitude and significance of the peril is a necessary precondition for doing the deed." However, I do not consider the survival of Nature in this regard as a "peril." Instead, it is a *duty*.

Montaigne agreed with much of Jasper's thought. He wrote: "What a stupidity to torment ourselves about passing into exemption from all torment! Extinction likewise brings not suffering but the end of suffering...You are dead after life, but during life you are dying; and death affects the dying much more roughly than the dead, and more keenly and essentially than it does the non-living who are not the dead but the unborn...Like death, extinction is felt not when it arrives but beforehand, as a deep shadow across the whole of life." Finally, Pascal described perfectly our response to the peril of extinction: "We find it much easier to dig our own grave than to think about the fact that we are doing so."

Jaspers ended with criticism of the wisdom in establishing a world government to oversee the prevention of possible thermonuclear war through full nuclear disarmament because such a powerful central authority "itself might have to possess nuclear weapons, which can only increase misgivings about this institution."

Einstein had similar doubts: "The unleashed power of the atom has changed everything save our modes of thinking, and thus we drift toward unparalleled catastrophe."[77] I believe that Einstein's misgiving is unjustified by evidence from history. In World War Two, all major participating powers possessed huge stockpiles of

---

[77] Schell, Jonathan: *The Fate of the Earth*. II. *A Deliberate Policy*. Stanford Univ. Press, Stanford, CA, 2000, 97.

poison gas that were never touched. That multilateral possession coupled with the sobering knowledge gained from the prior widespread use and horrific effects of the weapon were an invincible reining force. Furthermore, the renegade Nazis who were most likely to resort to such use, desisted even when facing imminent and inevitable doom and despite the fact that they alone possessed a superior and far more lethal new type of nerve gas.

Although fighting a losing battle to the end, as the Nazis did, may seem to promise a more noble sort of massive self-suicide, the cost of a population's annihilation through the most gruesome and excruciating kind of agony nerve gas can wreak is a sufficient deterrence to prevent exactly what Einstein feared. And that same "brake" will be operative in any future deliberation over the likelihood of thermonuclear war.

Now, back to the matter at hand. The above fairly comprehensive consideration of the potential effects of our now ominously elephantine "People-Bubble" should give us reliable guidance in devising a sure plan at least to *deflate* it and watchfully keep it deflated. You'll note that simply stopping its expansion is not an option because as an expert has clearly stated, "More people, all demanding greater affluence, *have already overshot the earth's carrying capacity for Homo sapiens*" (emphasis mine). And that's a prescription for certain species suicide, perhaps mortally wounding Nature in the ugly process. Let's face it: the current size of the world's population somehow *must* be reduced. And that vexing necessity calls for honest, unbiased, brutally frank and open discussion involving all classes of every society—whether they want to or not.

There's no time now for leisurely experimentation by the so-called scientific method with its expected necessary and time-consuming "blind alleys." The only acceptable specification for any measure(s) to be implemented is that they be absolutely and certainly effective. So, brace yourself; here are my preliminary suggestions put on the table only for debate, not for promulgation. In essence, the Solution consists of the same measures for either the Mini or Mega sort, the difference depending upon the magnitude of the problem to be solved. Thus the aim of species suicide will necessitate far more Draconian measures to achieve an absolutely certain, complete, and irreversible end by prohibiting *every* thing or action that could in *any* way preserve, abet, or enhance human existence as now manifested on earth.

(1) The first step would seem to be the establishment of the authority to decide on the Solution to be implemented; control and supervision of its operation; and what, when, how and why changes are to be made. None of this must be the prerogative of *any* established professional or lay group or any public or private political, governmental, or sectarian organization or institution. Instead, such authority should be vested in a totally independent panel consisting of as few a number as possible of generally respected members of the public, certified professionally as being sane, and judged by adequate documentation of training and experience to be qualified and suitable to serve in separate special forums as representatives of the constituent populace in each continent of the globe.

The panel's first task will be to set the most desired limit of population size if its choice is to risk maintenance of permanent continuity and natural ecological balance of the species' perpetuation through global cooperation using the cyclical route.

That would require strict control of the birth rate by the populace of every continent, first to stop the rapidly accelerating population growth now underway and that at the outset of any plan will already be at an unsustainable level. Therefore, an enforceable "two children per family" policy, which may maintain that level, may be out of the question from the start. Then, secondly, the level must be reduced to one deemed to be possible to maintain. That calls for a possibly intermittent "one child per family" policy, by means of which undesirable or inadvertent fluctuations can be controlled.

The arithmetic is simple: at any given level, one child for every two parents may in time probably reduce the population to, or close to, the desired level (considered to be about two and one half children per family) in one or two generations. The workable formula for this suggestion, which I realize sounds like an impossible fantasy, but it is theoretically possible of being devised only through global participation of sincerely benevolent and committed human beings—which generally Homo sapiens are not. Nevertheless, I bring it up only as a clarion to "think before we think," as perchance nothing other than an acceptable token surrogate for a "collective sacrifice to futurity."

Forums must be established without codification by laws enacted by conventionally elected legislative bodies. Forums will function under their own regulations set down by the specially selected delegates' majority vote, simultaneously, individually as well as collectively within six months after the date designated for their initial convention. Their subsequent rulings regarding procedural matters and measures to be adopted or changed for population control will also require a simple majority vote.

(2) The world's total population will be permanently limited to near the amount estimated by Hoyle to be at or near the end of each *subsequent* cycle, or about two to three billion people. That corresponds to the recorded totals in the first cycle between the years 1927 and 1960, a period during which life was fairly

comfortable with the existence of enough labor-saving technology, an intact and effective educational system, and many other accouterments of an adequately civilized, less frantic life mode that would be the envy of Alley Oop's cohorts.

(3) The following measures will be instituted to preserve and simplify at least that level of lifestyle.

a. Any significant evidence of extraordinary increase of fertility will immediately result in the mandatory invocation of abortion until satisfactory reduction is accomplished; and if that proves to be ineffective, then mandatory infanticide and neonaticide will be invoked.

b. All neonates born of rape, incest, or casual encounters out of wedlock will be destroyed. Immediately after the designated population limit is significantly exceeded; and if exceeded to dangerous levels, *all* newborn infants will be promptly destroyed.

c. If repeated by one or both of the same "parents," those neonates also will be destroyed and the "parents" sterilized.

d. Adoptions will be prohibited and the unwanted infants destroyed.

e. All practice and research dealing with fertility and artificial technological innovations aimed to combat infertility will be prohibited.

f. All forms of surgical manipulation to facilitate or enable the process of parturition, including gynecological surgery as well as surgery to correct or modify neonatal birth defects, will be prohibited and the neonates destroyed. All hospital inpatient obstetrics and special premature birth wards and activities will be abandoned. Births will take place in homes or in some cases outpatient clinics without subsequent confinement.

g. Neonates born outside of hospitals, inpatient clinics, or homes will be destroyed, and repeat offenders will be harshly punished.
h. Placentas from hospital and clinic births will be made available gratis to all who desire and need such sustenance.
i. All medical and surgical procedures requiring more than 5 (five) full days of inpatient hospitalization will be prohibited.
j. Research in human immortality, or to discover or create new medicaments or new medical procedures, will be abolished.
k. Mortuary practice will be limited to immediate simple burial in the ground of the corpse in a mass unmarked grave without embalming or special procedures; memorial ceremonies must be privately arranged in the home (as was customary in the past). The manufacture and use of caskets, the construction of mausoleums and above-ground crypts, new cemeteries, pet cemeteries, and kennels will be prohibited.
l. The preparation, distribution, and sale of all pet food will be prohibited.

(4) The use of any animal in medical, surgical, and cosmetic research will be prohibited. All of that can be conveniently accomplished on the researchers themselves or on other eagerly altruistic or paid human volunteers who give their duly informed consent. I was gratified to learn that the Scottish biologist J. B. S. Haldane followed his father's example by experimenting on himself. He concluded that "one great advantage of working on oneself…is that far greater

accuracy is possible." Pierre Curie studied the harmful effects of X-rays on his own forearm.

Most of my research has been accomplished on myself or on willing volunteers with their informed consent. Only in one instance did I use a condemned dog to test a theory relating to restoration of coronary circulatory insufficiency—which failed due to my lack of training and of adequate resources and trained assistants. Other more fortunate instances involved the awarding of the Nobel Prize to a German physician who first developed successful cardiac catheterization on himself, an experiment which opened the whole field of cardiac surgery. And an Australian physician, who quite recently won the Prize by proving on himself that peptic ulcers are caused by an infectious microbe, putting some surgeons out of a job.

It is common knowledge that paid volunteers now comprise an unusual and special occupation by enrolling themselves in a variety of medical, surgical, and pharmaceutical research projects. Just imagine how many animals are thereby lucky enough to evade a torturous and often grotesque death! Schopenhauer was absolutely right about human nature.

(5) College and university scholarships and parochial, private "charter" and "magnet" schools, and school busing and school luncheons all will be abolished. An intact Nature's survival is paramount. We just don't need more innovation that not only increases unemployment, but also fosters an increasing convenience, comfort, and boredom of life, permitting us to delude ourselves with the degrading amusements, fads, and idiocy that we crave to fill the spiritual vacuum eroding our lives—especially tragic for the innocently "dumbed-down" youth who can't even speak, write, spell, or communicate decently without the most vulgar and repetitive expletives. The

renowned 17th century author, Alexander Pope, summed it up simply: "Amusement is the happiness of those who cannot think."[78] He could have ended the statement, quite justifiably, with this: "or those who are trained how *not* to think," which would be a more accurate assessment of the effectiveness of our current public educational bureaucracy.

(6) Based on my experience with traditional education in America, from kindergarten all the way up through a top medical school, I have come to the conclusion that the present educational system should be abandoned entirely in favor of establishing the much more promising "deschooling" approach that does away with the ineffective formalities of forced attendance in traditional orderly classrooms, providing "jobs" for university trained "licensed" teaching staffs, top-heavy administrations, standardized curricula that are unsuitable for every student, and the "incarceration" of often bored students stripped of their natural rights for most of each day and year.

Deschooling, which I believe is the best way to discover and nurture latent talent and genius, is closely akin to the old apprentice system. It encourages the best education by teaching oneself through hands-on as well as question and answer sessions on subjects of interest involving extensive informal interaction with other similarly motivated students who thereby want to "teach" each other. I prefer that stimulating, casual, and very instructive method of general education, which I unconsciously had followed on my own for much of my adolescence and early adulthood in several diverse disciplines of special interest to me. According to the

---

[78] Brussell, Eugene E.: *Webster's New Dictionary of Quotable Definitions*. Prentice Hall, Englewood Cliffs, NJ, 1988.

noted educator Ivan Illich, "Deschooling accomplishes much more while eliminating the needed public support through a regressive taxation to cover the high costs of a huge professional apparatus of (sometimes incompetent and ill-trained) educators and buildings."[79] It certainly can't be any worse than what we have now.

All of the above suggestions may strike many readers as unreasonably harsh. However, reversion to a lifestyle comparable to that during the interval between the years 1927 and 1960 shouldn't seem so bad, despite the intermittent terrible occurrences of preventable events such as war and economic upheavals. The latter would be less likely foisted on a "more intelligent and sociable" populace less prone to allow those with base human natures maliciously to manipulate at will all industry and commerce. As stated before, in those days, there seemed to be enough social stability to establish and preserve family life and nurture human beings with the self-discipline and sense of responsibility that fostered and preserved optimism for the future, and for looking forward to experiencing it, free of the anxiety and fear of being deprived of it all. That is my brief analysis, based on 81 years of having observed the ever-changing human predicament.

Massive overpopulation isn't the only problem necessitating my harsh suggestions. The famous social anthropologist, William Graham Sumner, warned us in the early 20th century that humankind may find itself "between a rock and a hard place," as though squeezed in a sort of social "vice." In his opinion, "Where the earth is under-populated and there is an economic demand for men,

[79] Illich, Ivan: *Deschooling Society*. Harper & Row, NY, 1971.

democracy is inevitable." Such was the case when our country was born and in many respects also when the world's population was about two billion, as in 1927, and about three billion in 1960. "That state of things," he continued, "cannot be permanent. Therefore, democracy cannot last. It contains no absolute and *eternal* truths." And that appears to have happened already in the example of our own diminishing democracy, as discussed above. So, once achieved, the two or three billion level would seem to be easier to maintain when the harsh suggestions are operative.

Now Sumner's assertion causes us to take the wise course of watching very closely the attaining and preservation of those levels, to keep track of their fluctuations, and to be alert to detect any exaggerated deviations that could herald a disastrous change of population level in either direction, up, to anarchy and chaos, or down, to tyranny—all of which seem to complicate what at first appeared to be a simple theory for population control. No one can predict how it will work out in practice; but it probably won't matter because most of us doubt that humans either really can or desire, to undertake such a fateful challenge that involves drastic changes.

The last doubt is even stronger when it comes to the Mega-Solution, which probably every reader would deem to be absolutely impossible to implement. But it really isn't a situation essentially much different than that faced by Alley Oop and Oona. They had to ponder the "absolutely impossible" implementation of some kind of measure to reduce and keep level their community's basically sparse but circumstantially excessive number of people. They dared to do the "impossible" solution required by turning to the certainty, for them, of some sort of repugnant and unthinkable measure, which happened to be infanticide.

Did their hard reality make them spiritually stronger than our comfortable one that causes most of us to wince when merely hearing mention of the requisite measures? Or was it because their probably less sophisticated mythology had not yet petrified into a stifling creed? Nevertheless, the difference, no matter how small, is a clear sign of our civilization's divine mythology having successfully completed the atrophy and degeneration of our species' *natural* morality. That fatal derailment in the course of cultural evolution transformed, as if by magic, a naïve, so-called sapient being into another definitely worthy of the moniker "Sap." He failed to heed the following wisdom from a few of his genuinely sapient cohorts whom he cursed as "heretics" and desired to "rehabilitate" (preferably by roasting them alive)—the wisdom that:

Your ethical codes *must* dispense
With *all* dogma purblindly intense.
Without going to church you
Can manifest virtue
Through reason and plain common sense!

The putative measures for the absolute, irreversible elimination of even the possibility of ominous variations in population levels are ridiculously easy to enumerate, and they will indubitably guarantee the rapid reanimation of an intact, now unhindered Nature to maintain a stable, perpetually "democratic" ecological balance among its biological constituents—all accomplished by the vanishing of just one responsible species, Homo sapiens.

Let's consider what it will take to achieve that aim. It is easy to do because the measures needed to be taken must simply derange or eliminate the metabolic processes that preserve human's corporal integrity—or simply kill him. Therefore, all medical, dental, and pharmaceutical facilities, training, practice, and research must be ended. This includes hospices, nurseries, day care centers, and old age and nursing homes. Euthanasia and assisted as well as all other kind of suicide must be legitimized and encouraged. Vaccination and vaccine research must stop. And health insurance should be ended. Vasectomy of all males may become mandatory and pregnancy criminalized in especially exigent crises. Any infraction of the rules will be severely punished, and repeated infraction by an individual will result in his or her superfluous and uncontrollable life being ended by execution.

All forms of public and private charity must be abolished, including governmental subsidies such as Social Security, unemployment insurance, and food stamps. Homeless shelters and soup kitchens, too, must be abolished. Crime control will disintegrate, and prison riots will be rampant and almost impossible to control, unless incorrigible inmates are eliminated. All sorts of cannibalism will be absolutely prohibited, but probably will be permitted and commonly practiced nevertheless during random occurrence of intermittent crises of overpopulation.

These excessively harsh measures will become meaningless when the last human being has disappeared. Human extinction will then have "wiped the slate clean" to give Nature another opportunity to try evolving something better the next time by paying extraordinary attention to the quality of character it instills into that "something" if it is meant to be another and much improved, *really* sapient type of Homo. Nature and the world have had enough of the SAP variety.

Epilogue I

Life as a short course of twists, curves and bends,
Is richly replete as it sinuously wends
With much joy, pain, and strife
Through this maze we call life
To the void of extinction that waits at both ends.

Epilogue II

I decided that it's appropriate to append some brief personal comments of my own, on a few of the assertions and opinions of Jonathan Schell, the author of one of the most interesting books I have referred to in this presentation. (*The Fate of the Earth, part 2*: *The Second Death*, Stanford University press, 2000.)

Page 99:

> "To save humanity from extinction by nuclear arms...a good first step would be to order the destruction of all the nuclear weapons in the world...A logical second step would be to destroy the factories that make the weapons."

Schell seems to be almost obsessed with the *probability* of such a nuclear "holocaust." We already have evidence that a nuclear bomb explosion does not kill every person in the immediate vicinity or even at the center of such an explosion. And any extinction caused by such a global war would be indirect and protracted due to massive or complete nullification of an indispensable necessity of life, Nature. Doing away with all nuclear weapons and factories would supplement, reinforce, or merely retard such nullification through the unnecessary waste incurred in the pursuit of such endeavors. Furthermore, the result of a much more likely nuclear power plant accident will certainly cause widespread death and destruction locally and perhaps indirectly even globally—but probably not total extinction of the human species. Finally, the "people–

bubble" will accomplish the latter certainly and fairly quickly without any additional harm to Nature other than that ensuing from the horrendous rapine associated with conditions of complete anarchy when the end of our species is accomplished.

> Page 107:
>
> "In the last analysis it is all of mankind that threatens mankind."

He's absolutely right on that point. Also, extinction through the "bubble" is more moral, completely natural, whereas a nuclear debacle is completely artificial and *purposefully* contrived.

> Page 128:
>
> "In extinction a darkness falls over the world not because the lights have gone out but because the eyes that behold the light have closed."

There is no way for us to know the real essence and meaning of individual and collective extinction; all we can know is the permanent absence of a species collectively and the absence of some sort of indefinable and inscrutable "spiritual" essence individually. When the lights go out for an individual, it can only be in connection with dreamless sleep because in that case we surmise that the subject is experiencing total unconsciousness, which rules out any perception of darkness or pitch black. The latter represents "something," i.e. a shade of color, capable of being perceived even in an unconscious state, as proved on awakening by reliable reports as well as personal experiences of dreaming. Thus, with such verification on awakening, a dormant consciousness is *assumed* to have existed even in a dreamlessly sleeping subject. Whereas in death and extinction, without that evidence, we "know" for sure that cannot be true in a subject whose consciousness *evidently* expires completely with the body that sustains it and thus may not exist like everything else in the annihilation. To that extent, the three apparently similar entities of dreamless sleep, death, and extinction seem to be radically dissimilar.

In this regard, three other kinds of non-being deserve somewhat more consideration than our merely naming them. One is the status of the *unborn*, of not ever having been born. As far as we *can* know, it is no different than extinction per se. However, the latter concerns a being previously known to have palpably existed and whose non-being is therefore *real*, whereas the unborn's state of extinction must be *virtual*. Another kind might best be called "temporary extinction" such as the illusory *resurrection* of Christian mythology, which concerns the expected intact "restoration" on earth, or in some other imagined site, of a previously extant being. And finally, the equally delusional, occasionally sanctified "temporary" extinction known familiarly as *reincarnation* of a specific, intact, previously extant human being or one transformed, involuntarily or by choice, somewhere, into another live member of an unknown, fervently desired species.

The utter lunacy of these delusions, especially with regard to reincarnation, prompts me to offer this disrespectful criticism:

Reincarnation's a big hoax

That's worshipped by "*karma*-nized" folks

Whose hopes that it's true

A frog will undo

By not changing whenever it croaks.

Page 130:

> "There can be no justification for extinguishing mankind and therefore no justification for any nation ever to push the world into nuclear hostilities which...may lead uncontrollably to a full-scale holocaust and to extinction."

I was a little surprised and disappointed to read that Schell believes that. I assume that he means collective extinction. Evidently, it's only a matter of degree because it is accomplished by every war, perhaps not on a species but rather on scattered individual and group basis. However, Karl Jaspers undermines Schell's argument with the justification that "extinction could be necessary only as a sacrifice for the sake of eternity."

Page 133:

> Jaspers asked, "Is it possible, for example, that one day it might be our obligation to set both human interests and human morality aside and destroy the species for the sake of God?"

I answer: theoretically, "yes," on the slimmest of chances only for truly committed *religious* individuals unusually gifted with the extremely rare super-virtues of unwavering integrity, honesty, and self-discipline needed to fulfill such a weighty, collective, and somehow *religious obligation*, such as species martyrdom. For non-believers, the same is true with regard to an obligation, not for the sake of a purely fictive or mythological "god" or other imagined divinity, but rather for the sake of an existing real, palpable, terrestrial Nature—which is the viewpoint I express in this book.

Please understand that I am not foolish enough, and do not intend, to promulgate this issue as a "cause" or crusade which is an impossibility for Homo sapiens. My sole aim is to stimulate or, better yet, to *provoke* human beings into thinking about an issue involving everyone—a profound taboo too long neglected out of fear of ridicule or of just plain vacuous sentimentality over their beloved species.

I have long abandoned sentimentality at the risk of appearing to be somewhat cruel in debates and have staked my understanding of our empirical world with its nurturing Nature only on cold reason and common sense. I'll never forget the sage words of psychiatrist-philosopher Karl Jung: "Sentimentality is a superstructure covering brutality"—which probably accounts for most, if not all, of the terrible atrocities perpetrated in the name of religion throughout history. What else has the power to coerce human beings mistakenly and pseudo-piously to roast their kind alive and to discuss openly and pseudo-solemnly at the highest societal levels the advisability and "best" means of torture to apply to their kind?

Nature is accurately characterized by an anonymous proverb as a "hanging judge." She must have a somewhat dark sense of humor or amusement in having participated in the creation of humans and endowing them with an unalterable inborn nature that makes our future achievements predictable within the conditions Nature set. They were summarized briefly by the arch-atheist and attorney, Robert G. Ingersoll: "In Nature there are neither rewards nor punishments; there are consequences." Among the latter is the worst of all for Homo sapiens: our species' self-suicide as an inevitable consequence of our inability to renounce or at least control the sex drive, succumbing perpetually to Nature's devilish trick of having implanted in him the ultimate trigger for his annihilation: that excessive vigor of his pleasurable, irresistible, and invincible drive for reproduction.

It could happen soon, as an immediate event, within a few decades through human's volition to make this, our first of Hoyle's cycles, a futilely endured desperate effort to try to prolong the process of selfishly preserving the continuity of his coveted existence. Or, we may be able, at similarly great cost and inconvenience to ourselves to choose to even extend the agonizing desperation through the diminishing efficacy of a series of such cycles (as discussed earlier), which is most likely his preferred choice.

Nevertheless, as Sap thus agonizes, he may perchance begin to sense the significance of what is occurring and, remorsefully even, to feel slight pangs of sympathy for a long beleaguered, mercilessly smitten Nature and its—as well as his own—sadly abused domain. And, perhaps there's even an outside, though highly unlikely, chance that the pangs may be acute enough to enlighten and comfort Sap with the knowledge that the real significance of his sacrifice in thus dying is at the same time being transformed automatically into the nobility of a magnificent "beau geste" for futurity. That qualifies as an incontestable validation of a firm foundation for his species' worthiness of its extremely variegated, tumultuous, and sadly fleeting existence—a life that in some hopelessly cryptic way surely *must* have been worth living.

The gravity of what looms mandates that one deal with it in the only way one possibly can, as suggested by the poet Stanislaus Lec: "To think before you think," which unfortunately some cannot do.

Obviously, this is not a self-help book in the conventional sense. But for me and perhaps a few others, it might help a little to gain the satisfaction of a perhaps prescient awareness and certainly better understanding of, and readiness for, a more serene conciliation with our most probable and undeniably unsettling destiny.

Epilogue III

As we endure a sometimes seemingly endless eternity of travails that constitute the often difficult "obstacle course" of life, there is some degree of solace in the following thought that should never be forgotten:

Whether your life turns out nasty or nice
Will depend on fate's toss of the dice;
But you'll eventually find
That life's basically kind
Just by being *a merciful trice!*

Author Biography

Photograph by Amy Powers

Dr. Jack Kevorkian is a world-renowned person. He advocated terminally ill patients the right to die with dignity through physician-assisted suicide. Dr. Kevorkian was sentenced to jail for his convictions and served his term. He regularly appears at public speaking events. A movie, regarding his life, is being filmed with Al Pacino as Dr. Kevorkian, to be released in spring of 2010, and a documentary regarding him is to be shown in theaters in summer of 2010. *glimmerIQs*, an anthology of his writings, (which he composed while imprisoned), is published by World Audience, Inc. (2009).

CPSIA information can be obtained at www.ICGtesting.com
Printed in the USA
LVOW060425260613

340274LV00001B/61/P

9 781935 444916